Ctrl + Alt + Delete
Reboot Your Productivity

DONNA HANSON

Prime Solutions Training & Consulting Pty Ltd

First published in 2018

By Prime Solutions Training & Consulting Pty Ltd
© Copyright Donna Hanson, 2018

National Library of Australia Cataloguing-in-Publication Data:
ISBN 978-0-9775202-1-3

 A catalogue record for this book is available from the National Library of Australia

Cover, text design and layout by Almira Joy Bautista
Cover photo purchased under multi-use licence from www.canva.com
Author photo by Nancy Morrison
Editor: Stephanie Jaehrling

Disclaimer
All details in this book are current at the time of publication and are subject to change. Contents of this book are based on the use of Microsoft Office 2016.

All care has been taken in the preparation of the information herein, but no responsibility is accepted by the publisher or the author for any damage resulting from the misinterpretation of this work or from actions taken based on advice contained within this publication.

Trademarks
Trademarks have been omitted from the text for readability.
Microsoft Office, Outlook, Word and Excel are registered trademarks of Microsoft Corporation.

©2017 Google LLC, used with permission. Google and the Google logo are registered trademarks of Google LLC.

TABLE OF CONTENTS

INTRODUCTION

If you don't know where you're going, any road will get you there.

LEWIS CARROLL

Technology is vital in business today. We can't escape it.

Its 24/7 access has resulted in an increased workload, a sense of being overwhelmed, a loss of balance and blurred lines. It's a bit like parenting. There is no manual or rule book beyond the initial training we received when software was introduced. For many, we were simply given the computer and told, "Off you go!"

Occasionally, in performance appraisals, employees might have had a need identified. As a result, they were sent off to a classroom-style course that contained six to eight generic topics, with generally no relevance to the workplace. I know, because I used to facilitate these courses; staff would return to their office the next day with a day of work to catch up on, no context of where or how they could use what they had learnt and no time to try it out anyway.

After attending training, they were often regarded by their organisation as being "fixed", or as being the guru of that program. As if, like in the movie *The Matrix*, everything they may ever need (along with every possible context) was downloaded into their head in the six hours of training they attended. If only it was that easy!

Whilst it ticked a box, this type of training often offered little to no demonstrable improvement in skill, knowledge or productivity. For some, it was like a flashback to school and a less-than-inspiring experience.

Every person, team or organisation that I meet at conferences and events I speak at is looking for the magic bullet that's going to make their life easier with technology and enable them to get back to doing their jobs and living their lives.

Whilst I don't believe such a bullet exists, I'm hoping this book inspires you to *always* ask: am I working *smarter* and not harder?

This book is for individuals and teams wanting tools, tips, shortcuts and strategies to "reboot your productivity" – get stuff done, get back to work and enjoy life!

I think it's time to press Ctrl + Alt + Delete and reboot your productivity. What about you?

Donna Hanson

PART 1 – PREQUEL

If you're not getting better,
you're getting worse.

PAT RILEY, FORMER NBA BASKETBALL
COACH OF THE LA LAKERS

Before we can press Ctrl + Alt + Delete and reboot our productivity we need to be clear on a few things.

Firstly, we need to understand **why** we are where we are.

Whilst it won't change our current situation, knowing why we are where we are is a bit like when we make a mistake: it can offer us insights and lessons that enable us to see the signs so we can avoid reverting to the way things were.

Next, we need to look at the **what**. This shows what being where we are is costing us – physically, mentally and financially. Much of this is hidden costs, which we tend to assume is just part of business today.

Finally, we need to look at the **how**. How did we get here, what signs might we have missed along the way, and how do we ensure that we don't miss them next time, or in fact that there is no next time?

It's like leaving a trail of breadcrumbs when we know and understand:

1. why we are where we are (current situation);
2. what this means to us (consequences of not taking action); and
3. how we got here (how did we get to this point).

Once we identify this, we can wake ourselves up, smell the coffee, wine or roses (whatever your preference) and, should we head down that road again, recognise the signs and, more importantly, how to correct our course.

So let's get started …

CHAPTER 1 – WHY?

In the business world,
the rear view mirror is always
clearer than the windshield.

WARREN BUFFETT

Workplace expectations have always been subjective. When I first started work back in the late 1980s, the perception of "productivity" was measured by the time of day you arrived at work and the time you left.

Today it seems "productivity" is measured by the number of emails you send, who you send them to and when you send them. Both are perceived as this era's equivalent of a badge of honour, a measure of importance and productivity.

I once worked on a project for a global consultancy. Each month I facilitated an induction program for new starters. After my contribution to their professional education, the new starters were shipped overseas to various US, Asian or European locations. They were given a laptop and a smartphone and told to get to work!

There was no guarantee that they were productive; they were just taught how to use the tools and it was up to them to work out how this applied to their role and to getting their work done.

Today, unfortunately, "busyness" is still being viewed as a sign that we're being productive. But, really, often we aren't. Busyness doesn't automatically equate to productivity.

So, you're probably thinking to yourself: why now? Why do we need to change this?

Employees are suffering more from workplace stress due to work overload, increased expectations, corporate downsizing, and the pressure to get more done in the same amount of time, with the same or even fewer people and, of course, no extra tools or pay!

Medibank Private commissioned a report in 2008 titled *The Cost of Workplace Stress in Australia*. The statistics relating to health in the workplace are startling. Research shows the impact of workplace stress (which encompasses a multitude of elements) includes an average of 3.2 days per worker, per year in absenteeism and presenteeism (where workers are present but not productive), with a direct cost of over $10.11 billion dollars a year to business.

The human brain is wired for "fight or flight" for survival. Every piece of information we receive is processed and the brain has to determine what action to take. Before smartphones and email, the pace at which we received data was much slower than today. You could send a letter in the post and know you had at least a day or two before you would get a reply, either by telephone or return mail. Today the response is instantaneous.

The result is a constant bombardment of information that our brain is frantically trying to process. Although the brain is powerful, our mental processing speed is nowhere near that of a computer processor, nor will it ever be. The result is a sense of overload and increased stress.

With no universally accepted expectations or guidelines for using technology, humans have to create their own. It's like a noise in your car – you can pretend it doesn't exist, but eventually it will just get worse. The same applies to relationships – problems left unaddressed can cause them to fall apart.

These are pre-emptive signs that something needs to change. You may make the choice to change, but if you don't, you may find that the change is made for you in terms of your health or your stress levels.

About seven years ago, my husband hurt his back. He hadn't done anything specific that caused it. It seemed to get worse over time until he was in agony and couldn't move. We went to a surgeon who told us it was a bulging disc in his back: "The injury of the future", he called it. Our lifestyles and work modes are more sedentary than ever before, so unless we make changes to the way we work (such as using standing desks, doing eye exercises, drinking more water), we can expect back-related injuries to increase.

Poor eyesight is another example of the ill-health being caused as a result of increased work in front of computer screens – a boon for optometrists. Additionally, many people experience back and shoulder pain as a result of being hunched over at a screen all day, every day.

We can't continue to live like this. We can't continue to be overloaded. It's time to get back to basics, but before we can do that we need to understand why we need to change. Work is a part of our life, but it shouldn't define us. It is a mechanism for us to afford a lifestyle and pay for the things that we and our family want and love to do, but no amount of money can replace our personal health.

The next step is understanding what will happen if we don't change something.

CHAPTER 2 – WHAT THIS MEANS

Change your thoughts and you change your world.

NORMAN VINCENT PEALE

What if we don't do anything?

Technology was supposed to make like easier, but in fact it's created new problems, and it's snowballing. The result is increased stress, decreased productivity, decreased health, increased absenteeism and presenteeism (where people are at work but they're not mentally present or they are not productive at work).

To prepare for the future, we first have to look at how we deal with the present. If we don't change how we work, respond to and cope with things now, we'll get to a point of feeling overwhelmed and will suffer loss of health. How soon this might happen depends on how invested you are in your role.

Every time I present, I hear people say, "If only I'd known that last week", when I demonstrate a tip or a shortcut, how to quickly create a graph in Excel or how to sort and filter data easily or create a pivot table. I get comments like, "Wow, it's so easy", or "Five minutes instead of five hours".

The challenge with ease is the assumption of simplicity. Whilst a graph may seem to take only seconds to do, it took much longer to learn how to take seconds to do it.

One of my mentors, Million Dollar Consultant Alan Weiss, says, "You can always make another dollar, but you cannot make another minute".

We all have 24 hours in a day, but some of us seem to get more done than others. Why is that? It's because many of us let our time evaporate every single day because "You don't know what you don't know".

Did you know that if:

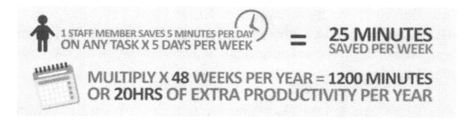

It has taken years for us to get to this point, so to think there is a magic bullet is absurd. Just like dieting and exercise, it takes time to get a result, but if we don't take action to change the way we work, the situation will only get worse and eventually impact on our physical and mental health.

Researchers at the University of London found that when we're interrupted by something such as an email pinging into our inbox and the message notification box appearing, it takes approximately 20 minutes for our brain to get back to the point where it was before the interruption happened.

Microsoft Canada conducted a Consumer Insights report on attention spans which revealed in the year 2000, the average human attention span was 12 seconds. In 2013 that had reduced to eight seconds.

This is startling, given the average attention span of a goldfish is nine seconds!

So what will happen to you and your family if you just keep doing the same things the same way and expecting different results?

CHAPTER 3 –
MOVING FORWARD

*The starting point of all
achievement is desire.*

NAPOLEON HILL

Today we spend more time planning our holidays than considering spreadsheets, documents or emails. In the real world we're short on time and as a result we often just throw together Excel spreadsheets and Word documents, hoping for the best and thinking things will sort themselves out.

Graduates and new hires may have the technical knowledge, or in an interview say that they do, but when it comes time to do the real work they can't make the connection between what they've learned and how it applies in the real world.

I once received a call from a global brand which needed assistance with a legal contract in Microsoft Word. The contract had initially started out as a few pages, but with the evolution of the business and the changing legal landscape, pieces were added over time and a table of contents was manually created. The document also had manually created multi-level numbering, i.e., 1.1, 1.1.1, 1.1.1.1, etc.

By the time they came to me, the contract was over 100 pages and they had been struggling for some time. A simple addition to a clause meant manually renumbering items and editing the table of contents, which was not only time-consuming, but posed significant risk to the integrity of the contract. Even more so, given the fact it was a legal document!

Often people say to me in workshops, "we just do a Save As and we change the information so it reflects what we want". Whenever I hear this, I cringe at the thought of important data being accidently overwritten, and the potential risks of making quick changes where we don't have time to consider whether the change is needed or whether we've missed something, increasing the likelihood of errors, omissions and unprofessional impressions.

In pre-computer days, a letter would be typed, then corrections would be made, and then it would be retyped, edited and fine-tuned before it was sent. Once sent, there was a window of a day or so for it to reach the receiver, who would then either call to discuss its contents or write a letter back, resulting in a lag of potentially two to three days.

With the instantaneous nature of technology today, we have fast-tracked the processing of data, saving us time, but we seem to have lost some of the care associated with communication, resulting in overloaded inboxes, documentation needing reworking and spread-sheets being inaccurate.

It's important to consider how you use the programs you use, as well as why you use them and what you use them for. Have you been shown how to use a program by somebody else? Are you self-taught? Did you attend a course? Maybe it was when you were at university, and you were just shown stuff and had no context, much like the graduates mentioned previously.

Many people will create an Excel spreadsheet, enter data and pass it on to someone else. The next person will enter some data in or make changes. Before you know it, mistakes have happened and formulas are incorrect and as a result data is wrong.

The common reaction is finger-pointing, but the reality is that in our busy world no one has the time to go through an Excel spreadsheet and check every single formula. What we need to do instead is set up our documentation with the goal of securing our data to minimise risk, simplify input and create efficiencies.

Do we just want to throw something together that will serve a purpose? Whilst that might solve the problem right now, it doesn't solve the problem in the future. In fact, like the global client in the earlier example, it can make the output even worse (and add lots of time to fix it).

Note: With the global client, there were two choices. Recreate the document from scratch with the correct headings and numbering setup (the easiest and quickest route), or painstakingly edit the existing documentation. Take a guess which choice they made?

So, why is it that we spend more time planning a holiday than planning how we will write a letter or produce a spreadsheet that we, along with others, will rely on for information and decision-making?

The plan with this book is to develop a strategy that works for you, and in the next few chapters I'll take you through a range of programs used every day by people in business, and strategies and ideas to truly capitalise on their functionality, and create processes and habits to reboot your productivity.

We need to stop and plan for what our desired outcomes are. Former English Prime Minister Winston Churchill is credited with saying: "If you fail to plan, you are planning to fail"! It's important that we plan: plan the work we need to do, the time it will realistically take to do it and the resources we need to complete it. Having a plan and working to the plan will enable us to move forward and create processes and habits to ensure that we are more productive.

There are great benefits to planning. Five minutes saved each workday equates to two-and-a-half days per year. Turn that around to time lost and multiply it by a number of tasks and staff and, before you know it, hundreds of thousands of dollars in productivity have been lost, and stress and frustration have increased. The benefits of having clear outcomes are that you save time, and extra time results in increasing your productivity and reducing your stress.

Just to be clear, I'm not suggesting that you don't take breaks. What I am suggesting is that your breaks are purposeful and with intent. For example, a lunch break. It's a great idea to leave your desk to eat and get some fresh air, or get a change of scenery, to reboot and recharge, press Ctrl + Alt + Delete, and then come back mentally fresh so that you're ready to hit the remainder of the day.

Recently I worked with a retail firm. One of their team members regularly exports data from a business-specific system; she needs to tidy it up, remove extra spaces, sort and filter in Excel, remove some data and set it up in such a way that she can pass the report on to someone else for them to work with. However, she worked in an open plan workspace, which meant that every time she did this task, if she got interrupted, she'd have to start again. We discussed the outcomes that she wanted to achieve and within an hour we had created a macro, which, put simply, is a shortcut way of having Excel repeat an automated process for you. This reduces risk and increases efficiency because if the macro is right the first time, it will repeat the same process with the click of a button.

The macro we created involved a process that would normally take her a couple of hours twice a week. We invested an hour to work through what she needed and, as a result, with the press of a button, that task, which was taking her two hours every single time, was done in a matter of seconds. Do you think she was happy? She was ecstatic! All of a sudden, she could *understand* macros, a topic considered advanced, and she could see how it was going to make her life easier. In my mind that is exactly what technology is supposed to do.

It doesn't matter what level of skill you are at or you feel you are at: introductory, intermediate, advanced. I believe if you can see how something is going to be relevant to your world and save you time, stress and frustration, then you will understand it and you will truly "get it".

As part of this book's resources, I have built a dedicated online resource, which can be found at **https://tinyurl.com/DonnaReboot** (coupon code for complimentary registration is REBOOT). All you need to do to unlock the resources is to create an account and you can go there at any time. It's free and contains a whole range of video clips and checklists on how to be more productive in all the programs that we're talking about in this book. I've set up the online resource for you to return to, because you won't remember everything from this book, nor will everything be applicable.

Another benefit of planning and knowing the outcomes is that you will have more time available to you. More time, not necessarily to do more work, but to get some balance back in your life. Like being able to stop for lunch. Being able to leave work at five o'clock, if that's what time you are supposed to finish, or being able to go home and not have to do more work while your family is relaxing or sleeping.

An additional advantage of automation is it simplifies systems and processes and enables you to know there is less risk. When you set up your spreadsheets in such a way that the amount of human intervention is minimised, productivity goes through the roof because there is less need for error-checking. If you've done it right the first time, it's going to be right every single time. So what next?

Start to think about and identify the time vampires in your day as far as your computer programs are concerned.

- Is it email? Do you leave it open all day?
- Is it Excel? Is it checking formulas, or you just don't know what you don't know?
- Is it Word? Are you struggling to get that text that you've taken from one Word document into a new one and they have different typefaces.

Don't worry. I've got the answers!

By identifying your outcomes before we start, we can evaluate where we go from here. Once you identify your outcomes, think about how it would feel if you were less stressed, more in control and able to get on with enjoying your life. You may or may not love your job, but we usually need it to support our lifestyle. Our jobs shouldn't be our lives; they're a means to an end.

What are three outcomes you would love to get as a result of this book? Some examples are:

1. Feel like I have a plan
2. Have a framework for productivity
3. Tips for managing my spreadsheets, documents or emails

List them here:

1. _____

    _____

2. _____

    _____

3. _____

    _____

PART 2 – APPLICATION

Time is the scarcest resource
and unless it's managed nothing
else can be managed.

PETER DRUCKER

In the next few chapters, we'll explore four of the most common areas of frustration in business today. You may find you're in control of one and not others, or none at all. The point is to consider taking a different view of how you use programs at a strategic level to get better results and reduce risk, stress and frustration.

Each section has a range of ideas and opportunities for you to consider, and actions, items and opportunities to expand your knowledge.

Now it's time to turn the page and give yourself permission to press Ctrl + Alt + Delete and reboot your productivity.

CHAPTER 4 –
EXCELLING WITH EXCEL

---

*Do not go where the path may
lead, go instead where there is
no path and leave a trail.*

RALPH WALDO EMERSON

My belief is, you don't know what you don't know. Let's go about changing this.

I regularly speak at industry events, in-house training sessions and conferences about productivity with Excel. I focus on the fact that one thing alone could save you hours of time. Not just the initial time to do something, but the failure time associated with having to re-do things; whether they're things that you or a colleague has done. There's a cost to fixing things that are wrong.

A client of mine puts together a budget once a year that is sent out to various divisions of the organisation. The challenge for him is, when the budgets come back, people have changed various titles, deleted elements of the budget that weren't relevant to them, and renamed things. This results in additional stress and loss of productivity because he has to spend so much time doing failure work.

So what is the answer?

This client was caught up in busyness. He needed to take some time out with someone to consider other ideas and options that weren't obvious to him, because he was so close and emotionally connected (in the form of time pressure) to the data. It was a matter of sitting down, having a conversation and highlighting some key things that he could change to set up his data to prevent users from doing the things that were costing him lots of time and frustration.

Implementing this process resulted in reducing the time taken to work on the budgets every year from several weeks down to a few hours. Those are the sorts of time-savings you can make on an ongoing basis if you choose to look at things differently.

Sometimes, as they say, you can't see the forest for the trees. And often you know the stuff already, but you just haven't put the pieces of the puzzle together in the right way to get what you need.

In this chapter we're going to explore some opportunities for you to press Ctrl + Alt + Delete and reboot your productivity with Excel. So let's get started.

Shortcuts

Having a repertoire of ten to twenty "go to" shortcuts in Excel is priceless. It doesn't necessarily have to be keyboard shortcuts, but common shortcuts.

I'm regularly surprised at presentations and events that I speak at when people come up to me afterwards and say, "I consider myself an advanced Excel user and there were things that I didn't know".

We don't know what we don't know. This chapter is about changing that with Excel.

Below I've listed a few of my favourite shortcuts, but you can go to **https://tinyurl.com/DonnaReboot** (coupon code for complimentary registration is REBOOT) to unlock the extra resources.

Copy and paste data in Excel or Word with the right result

Depending on where you are and what you are copying you may find over ten different options for copying and pasting. Choosing the wrong one can leave you tearing your hair out in frustration, but a couple of seconds choosing the right one can save you hours of editing and leave you with hair intact!

Copy and paste options – Microsoft Word

Copy and paste is considered a simple option, however, there can be multiple options. It is important to understand the differences between them so you can select the option that best suits your needs.

Copy as normal using the copy icon or Ctrl + C.

When pasting, by clicking the drop arrow below the Paste icon, a variety of options will appear.

Keep Source Formatting

Selecting this option means the formatting of the document the data originally came from is brought over to the new document, so if the typeface or font size are different, the text will look different.

Merge Formatting

This merges the two together. The results of this will vary depending on the attributes of the text. When you rest your mouse pointer on this option you will see a "live preview" of what it will look like.

Picture

This option inserts the data as a picture which means it cannot be edited.

Text Only

This strips out the formatting such as colours, italics, etc., and places the text only in the destination location.

Choosing the right option can save you lots of time and frustration with having to reformat documents.

Copy and paste options – Microsoft Excel

Copying and pasting data may seem like a simple task, but when you are copying and pasting
data from multiple files or locations, formatting,
etc., can be a problem. It is important to understand the options available.

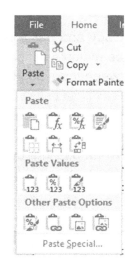

To copy data

Highlight data.
Press Ctrl + C or use the Copy icon. You can also right mouse-click and choose Copy.

Paste
Row 1
Icon 1 – Straight paste of cell contents

Icon 2 – Paste of the formula which will adjust according to the current location

Icon 3 – Paste the formula which will adjust the number format

Icon 4 – Keep source formatting

Row 2
Icon 1 – No borders

Icon 2 – Keep source column widths

Icon 3 – Transpose flips the data so items across columns shift to being in rows and items in rows shift to columns

Paste Values
Icon 1 – Value/result of the copied cell

Icon 2 – Value/result of the copied cell along with number formatting

Icon 3 – Value/result of the copied cell along with the formatting of the source cell

Other Paste Options
Icon 1 – Paste the cell formatting

Icon 2 – Link to the source cell – useful to ensure any changes automatically update

Icon 3 – Paste as a picture

Icon 4 – Paste as a picture with a link

Copy worksheets

If you have ever tried to copy and paste everything from one worksheet to another, you will know how frustrating it is to lose the column widths, shading, etc., when you do a standard copy and paste.

Instead of trying to copy and paste, click on the worksheet tab containing the data you want to copy.

per	Region 1	Region 2	Region 3	Combined Regions DC
ɔ2ſ U.UU	24ſ U4ſ ZUUɔ	I IUZɔ		
1,966.81	20/04/2005	11024		

Hold down the Ctrl key on the keyboard and "drag" the tab to the right. Let go of the mouse then let go of the Ctrl key.

The worksheet, including formatting, will be duplicated. (If nothing seems to have happened, it is likely that you let go of the Ctrl key before releasing the mouse.)

Check out the Copying Worksheets video in the resource area for full details and instructions.

Create graphs

Often you will need to create a graph when you are on a deadline. If you don't know what to do, it can be stressful trying to get it done, knowing you need to print it before you leave for a meeting in under five minutes!

Knowing how to create a quick graph with a simple keystroke will enable you to feel less stressed and more focused.

For a basic graph, simply highlight the data you want to chart and press the F11 (Function 11) key at the top of the keyboard.

Note: If you are using a laptop, to activate the F11, you may need to hold down an additional key such as fn or alt. Check the colour coding on your keyboard.

For a more specific chart, such as a range of non-adjacent cells, select the chart titles and matching data blocks (they must be the same size – check out the video for more details) using the Ctrl key on the keyboard (Ctrl enables you to select non-adjacent cells), then press the F11 key.

	Return To	Next			
Region 1					
Product	Government	<100 Seats	100-500 Seats	500> Seats	Total
Workstations	54000	23000	76000	175000	328000
POS Equipment	125000	45000	36000	110000	316000
Servers	75990	78000	17000	225000	395990
Software	24500	12000	24000	95000	155500
Total	279490	158000	153000	605000	1195490

Check out the Graph video in the resource area for full details and instructions.

Print titles

Titles enable you to get Excel to print out the column and/or row headings on each page, making it easier to read your data when printed out.

Click the Page Layout tab and choose Print Titles.

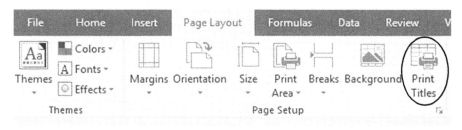

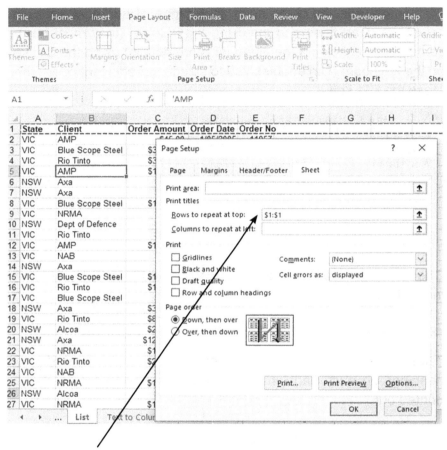

Click in the box beside "Rows to repeat at top" and then click on the row number you want to repeat on each page. If you want to repeat a column, click in the "Columns to repeat at left" and then select the column you want to repeat.

Note: You can choose more than one row or column to repeat.

Check out the Print titles video in the resource area for full details and instructions.

Hyperlinks

Most people are familiar with hyperlinks on the internet to navigate or move about web pages, but often haven't considered that they can be used to quickly and easily navigate around larger worksheets.

Hyperlinks are effective in Excel for linking from:

a) a worksheet/text/graphic to another in the current file
b) a worksheet/text/graphic to another file
c) a worksheet/text/graphic to another document, e.g., Microsoft Word document, PDF file
d) a worksheet/text/graphic to an internet page.

Start by highlighting the word, cell or picture that you want to turn into a hyperlink by clicking on it.

Click the Insert tab, choose Link.
Note: If the Link text is faded it means that you have not indicated what you want to hyperlink.

Link

Links

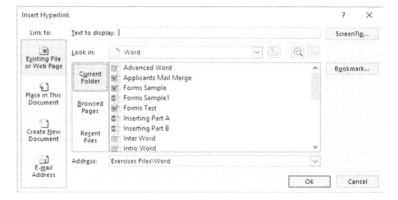

🖱 In the address box put either the file you want to link to or the URL of the web page.

🖱 If you choose Place in This Document on the left sidebar you will see a list of the worksheet in your file; just click on the one you want to link to create a link.

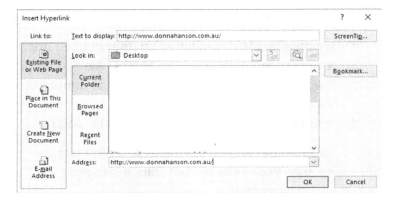

🖱 When finished, click OK.

🖱 A line should appear under your text and it will change colour to indicate that it is a link.

Check out the Hyperlinks video in the resource area for full details and instructions.

Range names

Similar to a hyperlink, range names enable you to navigate around a spreadsheet, but are even more useful as sometimes it is hard to understand a formula when it refers to a range of cells such as B12:K145. Range naming allows you to reference a range of cells by a title such as Sales.

To create a range name

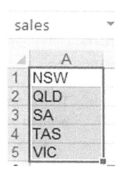

- 🖰 Highlight the cells you want to range name.
- 🖰 Click in the box to the left of the name box.
- 🖰 Type in what you would like to call the range. Note: 32 characters only, no spaces and it cannot look like a cell reference, e.g., A2.
- 🖰 Press Enter.

When you move anywhere in the file and you want to go back to your range name, simply click the drop arrow beside the name box near the formula bar and choose it from the list. Your cursor will automatically move to the range-named cells.

Range names make it easy to recognise errors in spreadsheet formulas. Rather than assuming a formula is correct, especially if others are able to edit the file, when a range name is used, you won't need to worry that a row or column reference is incorrect.

Select entire worksheet

Click at the start of the data on your worksheet (e.g., A1). Hold down Ctrl + Shift + End to select from the starting point to the end of text in the worksheet.

Note: If Excel seems to have selected blank columns or rows, it is because there was data in those cells at some stage.

Send a copy of a worksheet

There will be times where you want to send a single worksheet to someone.

🖰 The simplest way to do this is to right-click on the tab you want to send and choose Move or Copy.

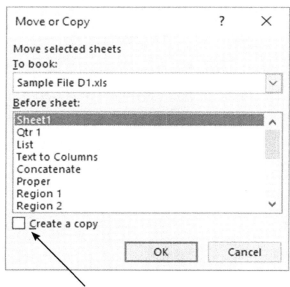

🖰 Click the checkbox "Create a copy".

🖰 Click the drop arrow beside the current file name and choose "new book".

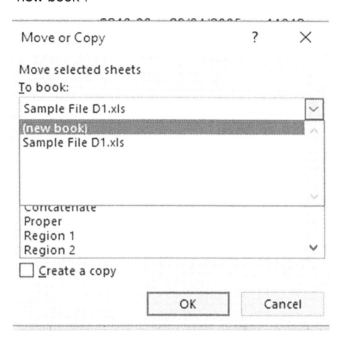

🖰 Click OK.
🖰 A new file will open with just the selected worksheet in it.
🖰 If you are using Microsoft Outlook for email you can quickly send it by clicking File Share.
🖰 Choose to share it as an attachment (in Excel format) or as a PDF.

Grouping worksheets

Grouping worksheets is useful if you have several worksheets in the one file that all have the same layout and you want to change the same item on all worksheets, for example, add in a new column or row, or type text in a cell that you want in the same cell on all the selected worksheets.

🖰 Click on one worksheet tab.

🖰 Hold down Ctrl and click on other tabs required (or hold down Shift and click the last tab required – Excel will select everything between those two points). Beside the file name at the top of the screen the word [Group] will display in square brackets.

🖰 Make the changes you want on this worksheet and the same changes, for example, text added in, rows added or deleted, etc., will be applied on all selected worksheets. The worksheets selected will appear white.

🖰 Once you have finished making the "group" changes, either right-click on another worksheet that isn't part of the "group" or right-click on one of the tabs and choose Ungroup.

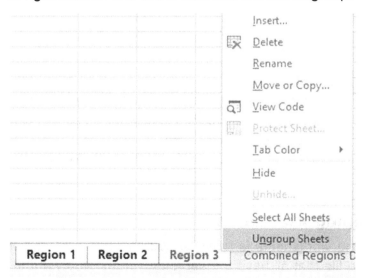

Transposing data

Sometimes in Excel you lay out a spreadsheet only to find that you need to "flip" the data, meaning you need to move the column headings and make them row headings and make the row headings column headings.

Rather than re-enter the data, Excel allows you to use a tool called Transpose to "flip" the data.

- Select the data you wish to transpose or flip.

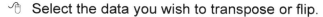

	A	B	C	D	E	F
1		Return To	Next			
2	Region 1					
3	Product	Government	<100 Seats	100-500 Seats	500> Seats	Total
4	Workstations	54000	23000	76000	175000	
5	POS Equipment	125000	45000	36000	110000	
6	Servers	75990	78000	17000	225000	
7	Software	24500	12000	24000	95000	
8						
9	Total					
10						

- From the Home tab, click Copy.
- Click where you want the "flipped" data displayed.
  Note: You cannot place the flipped data where the original data is in the initial paste. You need to flip it, then paste it to the old location.

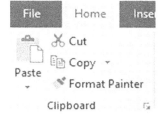

- Click the drop arrow below Paste and choose Transpose (it should be the last one in the first group).
- Click OK.
  Note: Data will now be flipped and placed in the new location. It can then be cut and pasted over the old location. Some reformatting may be necessary.

So what shortcuts will you utilise? I've covered some tips here but as a bonus, because you've purchased this book, you can go to **https:// tinyurl.com/DonnaReboot** (coupon code for complimentary registration is REBOOT) to access a range of videos clips on how to do these shortcuts in Excel.

Conversations

The next thing you need to do is start a conversation. This might sound a little strange; how can a conversation help reboot your productivity? Well, when you get together with your team, your department or your organisation, one of the challenges is knowledge-sharing.

The longer we work at an organisation, the more intuitive knowledge we build up. Because our knowledge builds up organically over time, we often assume that what we know is exactly the same as what other people know – which generally isn't the case.

Starting a conversation with your team, your department or in your organisation about how you're using Excel means you will start to know and understand the skill sets of your colleagues. You'll often find best practices exist in the brain of somebody else. In order to truly leverage our productivity, we've got to stop looking *outside* the organisation (e.g., Google) for the answers and start talking to our colleagues.

In the same way we talk about how we're going to deal with a particular customer, or how we're going to implement a particular project, we need to start sharing our Excel knowledge. We need to be comfortable with that and not feel like we need to house information in our mental silo that we never share with anyone – that's not contemporary thinking and reflects a scarcity mentality.

By choosing to be generous with what you know, your generosity will be rewarded by other people sharing what *they* know and we all start to make each other's lives easier. And wouldn't it be great if you were to raise a challenge that you're having with your team, your department or your organisation only to have someone say, "I've got an answer to that".

Think and Grow Rich author, Napoleon Hill, espouses the benefits of masterminds, which is simply a group of people getting together to share information for the betterment of each other. In effect, that's what this is. It's about creating a physical or virtual mastermind with your colleagues and sharing information that's relevant to Excel – talking with them about problems that you have and asking for advice on how to solve them, or sharing how you solved them. It can also fast-track your tasks, rather than you going off to try and find the answers, wasting valuable time and being stressed and frustrated.

By starting the conversation we're focusing on corporate knowledge. In the many organisations I work with, what generally happens is that everybody is busy doing their own thing, and it's not until somebody leaves that an organisation realises how much knowledge has been lost. If I had a dollar for every client who said to me, "We used to have someone here that knew how to do that but they've left and now nobody knows".

Unfortunately, it's often common things that have a lot of risk attached to them if mistakes are made, for example, a complex Excel worksheet with detailed formulas. With many intricate worksheets and charts, an errant keystroke is all it takes to turn a perfectly performing Excel worksheet into a "broken" one, or worse, you don't even realise there are errors in it! Often, all you need to do is ask a colleague to find out if they know a better way. It doesn't cost anything and takes very little time and, realistically, what's the worst that can happen? They say they don't have an answer and you really are no worse off. But what if they *do* know the answer?

If you need help with starting a conversation, download the conversations template from the resource at **https://tinyurl.com/DonnaReboot** (coupon code for complimentary registration is REBOOT).

Planning

The next thing you need to know about Excel is that you need to plan.

I had a client in the healthcare industry who had a lot of research data that they wanted to analyse. The problem was that they simply opened the Excel document and began entering the data *across* the worksheet. It was becoming very challenging to read. They realised it would be much better if the data went *down* rather than across the worksheet, but they didn't know how to change it, so they started recreating that worksheet with the data going down the page, resulting in a rework of many hours, plus the stress and frustration of thinking why didn't we do it the other way the first time!

For this client, rather than re-entering the data, all they needed to do was use the Transpose feature in Excel, as described above. This enabled them to flip the existing data over, so instead of it going horizontally it went vertically.

So before you start your next spreadsheet, take some time to plan. Ask yourself some questions before you start.

- What is the outcome that I want?
- What do I need this worksheet to do?
- Will others need to access this Excel worksheet?
- What's the risk associated with everybody having the ability to edit or manage this particular worksheet?

These are things that many people don't think about because they are often too busy ticking the boxes so that they can get stuff done, rather than taking time to make time, with the by-product of reducing risk.

Another question to ask yourself with every Excel worksheet you work on, regardless of whether it is something you or someone else has created, is: Can I assume this data is accurate? Every time someone else works on a document and it comes back to you, can you assume the data is correct? What we need to do is "lock it down" in some way to ensure that we know at every single point it is 100% accurate.

You need to ask yourself, what do you want your spreadsheet to do and where is the risk?

Proper planning decreases stress, decreases risk and increases productivity and the ease with which we do our work. The most important thing in life is our family and friends and our life outside of work. Our work is a means to an end, so we need to make it less stressful, so we can be as productive as possible to maximise our time at work, and then go home, switch off and be truly present with our families.

Automation

The next thing you need to know about Excel is automation. Whilst it may sound complicated, macros in Excel are a great way to automate some of those common tasks.

What would it mean for you with a task that you do repeatedly if, instead of having to do it manually every single time, you had a macro that enabled you to just click a button and have the job done for you? Wouldn't that relieve your stress and increase your productivity?

Macros

If you perform a task repeatedly in Microsoft Excel, you can auto-mate the task with a macro. A macro is a series of commands and functions that are stored in a Microsoft Visual Basic module and can be run whenever you need to perform the task.

Identify uses for macros

Put simply, macros automate a routine task. To work out whether a recorded macro will be of benefit or not, I like to ask three simple questions:

- Do you do something on a regular basis?
- Do you do exactly the same thing each time?
- Would it be easier to press a button to get it done?

If the answer to all three questions is yes, then a macro will be ben-eficial to you.

Record a macro

When you record a macro, Excel stores information about each step you take as you perform a series of commands. Excel stores this information in the Visual Basic Editor in a programming language called Visual Basic for Applications (VBA). We will record a macro that enters your first name in a cell.

- Click on the View tab, Macros and select Record Macro.
- Type in a name for your macro – no spaces and no charac-ters, up to 32 letters or numbers.

⌐ᗺ Choose to store it in the Personal Macro Workbook by clicking the drop list if you want to have access to this macro whenever you are using Excel on this computer.

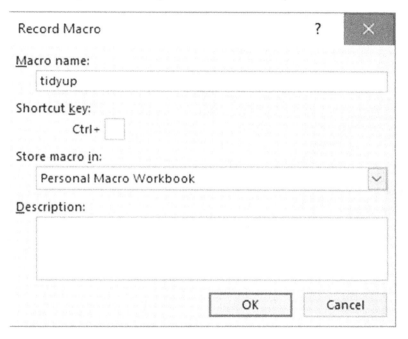

⌐ᗺ Press OK. You are now recording.

⌐ᗺ Perform your repetitive actions.

⌐ᗺ When you have finished, click View, Macros, Stop Recording.

Run a macro

You can run a macro by choosing it from the list in the Macro dialog box. To make a macro run whenever you click a particular button or press a particular key combination, you can assign the macro to a toolbar button, a keyboard shortcut, or a graphic object on a worksheet.

- From the View tab, click Macros, View Macros.
- A list of macros available in this file will display.

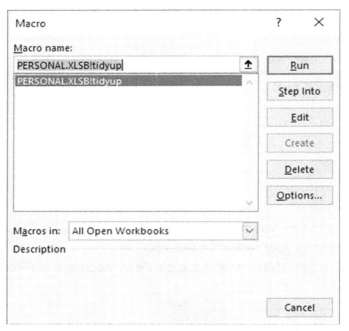

- Choose the desired macro and click Run.

Setting the macro as a Quick Access Toolbar button

- Click the drop arrow beside the Quick Access Toolbar.

- Select More Commands.
- Click the drop arrow beside "Choose commands from" and select Macros.
- Select the macros and click the Add button.
- Click OK to add your macro to the Quick Access Toolbar.

You will find a video of how to create macros in the resource at **https://tinyurl.com/DonnaReboot** (coupon code for complimentary registration is REBOOT).

The second way that you can automate, particularly if you're sharing a worksheet with somebody else for them to input or edit data, is to consider using something called a drop list. You've probably seen these if you've worked on the web or in other Excel documents, where you're on a cell and you click a little drop arrow and it shows you a range of options. These are relatively easy to do and are very useful for Excel worksheets where you need the data to be consistent.

For example, Vic, Victoria and VIC are three different types of entries in Excel, which, if all three are used interchangeably, can make it hard to analyse data. If you have your Excel spreadsheet set up with drop lists so that users can simply select from a list, you alleviate the problem of inconsistent data entry. This makes it easier for you to manipulate your data, and easier for people to input their data, because they don't have to worry about spelling; they simply select from a drop list.

Drop lists

A drop list is a form of data validation. It ensures consistency of data entered by users. For example, Excel recognises Vic or V or Victoria as three different entries even though they refer to the same thing. Drop lists enable users to select from a list rather than type in data.

To create a drop list

- Type the list of options you want users to have available (you may want to put this on a separate worksheet which you can hide to avoid users deleting them).
- Sort the cells A-Z by using the Filter & Sort button on the Home tab.
- Highlight the cells without headings.
- Click in the name box to the left of the formula bar and type in a range name. Note: A range name cannot have any spaces or symbols such as ?.
- Press Enter to turn the selected range of cells into the range name.
- Highlight Entry area, i.e., the column you want users to enter data into. Note: You can click the first cell and use Shift and the down arrows to select the entire column minus the heading.

🖱 Click the Data tab and choose Data Validation.

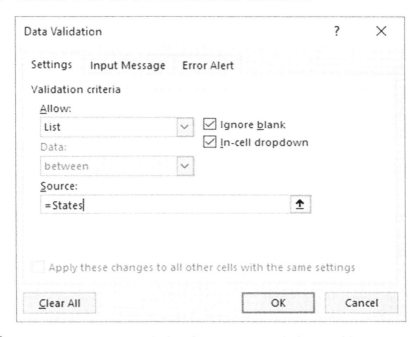

🖱 In the Allow field, click the drop arrow and choose List.
🖱 In the Source field type in the range name, i.e., =Range name.
🖱 Click OK.

When you click on a cell in the column a drop arrow will appear, enabling you to select from a list of options.

Think about your own work. Which of these items would deliver the most benefit to your productivity? Think about one thing right now which would be the priority item to help you.

What's your priority item in Microsoft Excel?

_____

_____

How could it transform your productivity and make your life so much easier?

_____

_____

You can access the Excel resources at **https://tinyurl.com/ DonnaReboot** (coupon code for complimentary registration is REBOOT) to see what could add value for you, and download the Productivity Planning Worksheet.

To summarise: In this chapter, we talked about the opportunities you have to reboot your productivity with Excel.

1. Learn some shortcuts.
2. Start a conversation with your colleagues.
3. Plan your worksheets; don't just jump in and start work – plan them.
4. Automate.

So, over to you. It's time for you to reboot your productivity with Excel.

CHAPTER 5 – WORD

Start as you mean to continue.

BILL GATES

Microsoft Word is often disregarded or overlooked as a problem area for organisations. I often hear: "We have Word under control. We don't have any problems with Word."

So, my question is, why do documents go out with misspellings? Why do they accidentally go out with the wrong clients' names in them?

We often don't think about needing to improve our use of Word because we assume we know how it works. It's simply like a typewriter, isn't it? No, it's not!

A document that has spelling or other errors in it, such as the wrong client name, is unprofessional, and it exposes you to clients who are making decisions about whether or not they want to work with you. You need to make sure that every document that leaves your organisation is a great first impression or a second impression or a third impression.

Back in the old days, when I first started work, I worked for an insurance company and we had a typing pool. We would write notes of what was to be typed up, put it in a little envelope and send it off to the typing pool. Miraculously, it would come back a day or two later fully typed up. We look at that today and laugh, because it simply wouldn't be acceptable in our world of 24/7 connectivity.

Those days are long gone. Now there is the expectation of autonomy. So it doesn't matter what role you're in. If you're in a sales role you need to know how to use Word for documentation, because if you need to redo a proposal five minutes before you walk into a client's site, you need to make sure that it's done correctly. You need to know what to do and you need to make sure that you feel comfortable and confident.

In a global market, with laptops, tablets and smartphones, we are able to create documents at the touch of a few buttons, so autonomy is so much more important. Let me give you an example of how it can impact on your productivity if you don't even *think* at the start about what it is that you're trying to achieve.

A client, mentioned earlier, had a legal contract that started with a single page, which expanded over time. It was gradually added to until it was 100 pages long. Everything in the document was manually numbered: 1, 1.1, 1.11, 2, etc., and they manually created a table of contents. The problem for the client was their table of contents.

Every time there was a legal change to the contract, they'd have to go through, edit the text, potentially change the numbering if a new clause had been added, and then reformat the table of contents. They were doing all of this manually, which was a huge time-waster. If you know your document is going to evolve over time, using a few tricks at the outset will save you hours. Start as you mean to continue.

In this section we are going to look at ways to reboot your productivity with Word, so let's press Ctrl + Alt + Delete. First we'll look at some shortcuts and how they can help you.

AutoCorrect

The first shortcut is AutoCorrect. This feature enables you to correct commonly misspelt words, automatically replacing them as you type. However, what can happen is that if you type ACN, which stands for Australian Company Number, and press the space bar it automatically corrects to CAN, C-A-N, and often that's not what you want. So you can also enter abbreviations for regularly inputted information.

Using AutoCorrect
- Click the File tab.
- Click the Options button at the bottom of the list.
- Click the Proofing option on the left and click AutoCorrect Options.
- In the Replace box, type in the abbreviation.
- In the Replace with box, type in what you would like to appear when you type the replace characters in.
- Click Add, then OK and OK again.

To activate, type in the abbreviation then press the space bar.

AutoCorrect is a very useful tool, not only because it corrects commonly misspelt words but because you can also add in your own shortcuts. So, for example, my name's Donna Hanson. I can type DH in my document and it will automatically expand my name. How many words do you constantly type that could be replaced with a quick AutoCorrect entry?

How do I do that? Go to the resources at **https://tinyurl.com/ DonnaReboot** (coupon code for complimentary registration is REBOOT) and take a look at the "how to" videos.

As I said in the last chapter, five minutes a day added up across a year equals two-and-a-half days!

Quick Parts

The next shortcut is Quick Parts. Quick Parts is a bit like AutoCorrect, except that you can use blocks of text. This is handy if you have clauses or standard blocks of text that you want to put in a document. You can even use it to create a whole document; when you open up a blank document or your letterhead, for example, you can just insert the Quick Parts and it will enable you to insert text exactly as you want it. Now, wouldn't that save you time if there are blocks of text that you use on a regular basis?

Using Quick Parts

- Type in the text required, e.g., a closing salutation.
- Select the text required.
- From the Insert tab, click Quick Parts.
- Choose the Save to Quick Parts Gallery option.
- Name the entry.
- Click Add.
- To place the building block or quick part into your document, go to the Insert tab, click Quick Parts and click on the desired block.

Go to the resources site at **https://tinyurl.com/DonnaReboot** (coupon code for complimentary registration is REBOOT) and you can see how Quick Parts works.

Templates

The next shortcut is templates. A template is a special file in Word that you use as the foundation for a new document. A template can be created for any document that is used on a regular basis, such as a fax header sheet or an electronic letterhead. For example, whatever size your business, you might have an electronic letterhead template with your details which you can insert and then create a new document from there. You might have a template for proposals where you work your way through the template, putting in the required details for a particular proposal.

The benefit of a template is that you don't have to locate a file, make changes and remember to do a Save As to save it as a new file. A template generally cannot be overwritten (unless you go specifically to the template file to change it).

Create a template

- ⌐ Open the file you want to save as a template.
- ⌐ Click the File tab and choose Save As.
- ⌐ Choose the desired destination.
- ⌐ Click the drop arrow beside Save As and choose Word Template.

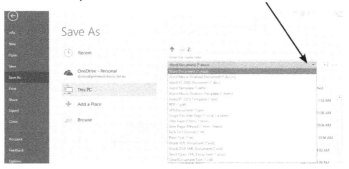

- ⌐ In the File Name box, type the name of your template, e.g., Letterhead.
- ⌐ Click Save.

Use the template

🖰 From the File tab, click New and Choose PERSONAL.

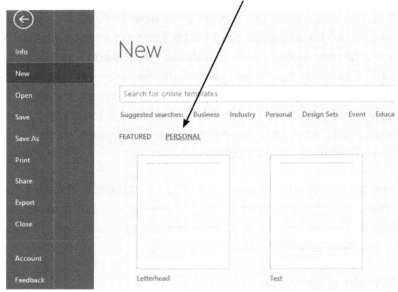

🖰 Double-click on the required template, e.g., Letterhead.
🖰 Edit the document and save it to the required file location.

Section breaks

The next shortcut is section breaks. Section breaks are handy for reports. They enable you to insert a landscape page into a document that is in portrait mode (vertical, the long side being the height of the page) and then revert to portrait as the document continues. This is useful, for example, when you want to insert a budget, but a portrait page can't contain the width of a budget. It enables you to create professional-looking documents, for example, for proposals which contain sales data or other specific information that all needs to be in the one document.

Insert a section break

- 🖱 Place your cursor where you want to "break" the page.
- 🖱 Click the Layout tab and choose Breaks.

I'll explain each of the options and an example of where you might use them.

Page break

This option allows you to insert a page break wherever your cursor is located. It enables you to start a new page where you want to. The keyboard shortcut for this is Control + Enter.

Column break

If you have a document with several columns in it, for example, a newsletter, column break ends the current column and places new text that you type at the top of the next column. This can be useful for "balancing" text that appears over several columns.

Text wrapping

If you have a photograph or image in a document and want to control how text wraps around the image you can use the text wrapping option.

Next page

Use this option if you want to have one page portrait and the next page landscape or you want to have different headers or footers in different parts of the document.

Continuous

This is useful for a document where you want to display text in one column (this is the default) and further down the same page you want another piece of text to display in three columns. See the example below.

> Video·provides·a·powerful·way·to·help·you·prove·your·point.··When·you·click·Online·Video,·you·can·
> paste·in·the·embed·code·for·the·video·you·want·to·add.··You·can·also·type·a·keyword·to·search·
> online·for·the·video·that·best·fits·your·document.¶
>
> To·make·your·document·look·professionally·produced,··Word·provides·header,·footer,·cover·page,·
> and·text·box·designs·that·complement·each·other.··For·example,·you·can·add·a·matching·cover·page,·
> header,·and·sidebar.··Click·Insert·and·then·choose·the·elements·you·want·from·the·different·
> galleries.·······················Section Break (Continuous)·······················
>
> | Themes·and·styles·also·help· keep·your·document· coordinated.··When·you· click·Design·and·choose·a· new·Theme,·the·pictures,· charts,·and·SmartArt· | For·example,·you·can·add·a· matching·cover·page,· header,·and·sidebar.··Click· Insert·and·then·choose·the· elements·you·want·from· the·different·galleries.¶ | Video·provides·a·powerful· way·to·help·you·prove·your· point.··When·you·click· Online·Video,·you·can·paste· in·the·embed·code·for·the· video·you·want·to·add.··You· |
> |---|---|---|

Note: When you delete a section break, the attributes of the section below are applied to the section above. See the example below.

Before

> paste·in·the·embed·code·for·the·video·you·want·to·add.··You·can·also·type·a·keyword·to·search·
> online·for·the·video·that·best·fits·your·document.¶
>
> To·make·your·document·look·professionally·produced,··Word·provides·header,·footer,·cover·page,·
> and·text·box·designs·that·complement·each·other.··For·example,·you·can·add·a·matching·cover·page,·
> header,·and·sidebar.··Click·Insert·and·then·choose·the·elements·you·want·from·the·different·
> galleries.·······················Section Break (Continuous)·······················
>
> | Themes·and·styles·also·help· keep·your·document· coordinated.··When·you· click·Design·and·choose·a· new·Theme,·the·pictures,· charts,·and·SmartArt· | For·example,·you·can·add·a· matching·cover·page,· header,·and·sidebar.··Click· Insert·and·then·choose·the· elements·you·want·from· the·different·galleries.¶ | Video·provides·a·powerful· way·to·help·you·prove·your· point.··When·you·click· Online·Video,·you·can·paste· in·the·embed·code·for·the· video·you·want·to·add.··You· |
> |---|---|---|

After

Video·provides·a·powerful· way·to·help·you·prove·your· point.·When·you·click· Online·Video,·you·can·paste· in·the·embed·code·for·the· video·you·want·to·add.·You· can·also·type·a·keyword·to· search·online·for·the·video· that·best·fits·your· document.¶ To·make·your·document·	point.·When·you·click· Online·Video,·you·can·paste· in·the·embed·code·for·the· video·you·want·to·add.·You· can·also·type·a·keyword·to· search·online·for·the·video· that·best·fits·your· document.¶ To·make·your·document· look·professionally· produced,·Word·provides·	text·you·want.·If·you·need· to·stop·reading·before·you· reach·the·end,·Word· remembers·where·you·left· off---even·on·another· device.¶ Video·provides·a·powerful· way·to·help·you·prove·your· point.·When·you·click· Online·Video,·you·can·paste· in·the·embed·code·for·the·

🖱 Click on the desired break.

Headers and footers

The next shortcut is headers and footers. A header is text that appears at the top of every page and a footer is text that appears at the bottom, for example, page numbers or titles. Longer documents often have a different header and footer on the first page compared to the subsequent pages, particularly reports, etc.

Headers and footers enable your documents to look professional and consistent, and also enable the reader to see what part of a document they're in. If a document is divided into sections, or you have a report where different people contribute data, you might want to separate the information and have different headers and footers for each section.

Create a header or footer

The quickest way to insert a header or footer in a document is to double-click at the top or bottom of the page.

This will activate the Header & Footer Tools Design tab.

- By default, the cursor will be placed in the header of the current page.
- Various options are available from the Header & Footer ribbon.

Options include: Page Number, Date & Time, Quick Parts, Picture, Clip Art, Go to Footer, Previous section, Next section, Different First Page, Different Odd & Even Pages, the distance Header/Footer is from top and bottom, Insert tab and Close.

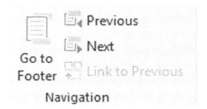

To move from the header to the footer click the Go to Footer button.

To switch the Header & Footer toolbar off and return to the document, click the Close button on the ribbon.

Different headers and footers in different sections

Once you have inserted section breaks into your document you can use different headers and footers for different sections. For example, Section 1 might relate to a chapter of a book and have a title relevant to that chapter; Section 2 might relate to the next chapter of the book, requiring a relevant title for that chapter.

To have different headers and footers in different sections of your document you need to first insert section breaks.

Activate the appropriate section header by clicking in the section, then clicking the Insert tab and choosing Header. Header and the section number will display as per the example below.

Whatever text or information is in the first section header will automatically display here. If you want different text to display in this section, de-activate the Link to Previous option. This will ensure the information you place in the header will not be applied to previous sections. You will need to de-activate the link in each section if you do not want the previous section's header and footer to repeat in the new section.

There is a video in the resource area on using section breaks combined with headers and footers, and how the two can interact with each other. Headers and footers are a quick shortcut that enable you to increase your productivity.

Remember, five minutes every day, 48 weeks of the year, equals two-and-a-half days. Every five minutes you save, and you save that every time you do that task, equals two-and-a-half days you've gotten back every single year.

Macros

The next shortcut is macros. I talked about macros in the Excel chapter, but in Word there are a number of different macros. Probably the most common one is a macro that creates a template document. For example, a macro that saves you going: File, New and choosing a template. You can create a macro so that you push the button and

Word automatically does all that for you. Or perhaps you have a document that you need to format. When you put a document together (it might be from multiple sources), you need a macro that goes through the entire document, does a spell-check, and applies your organisational typeface and font size.

Macros can be used to automate repetitive tasks by recording the steps involved. Macros can be stored in the personal macro workbook which makes them available whenever Word is open.

Create a macro
   ✍ From the Developer tab, click on Record Macro.

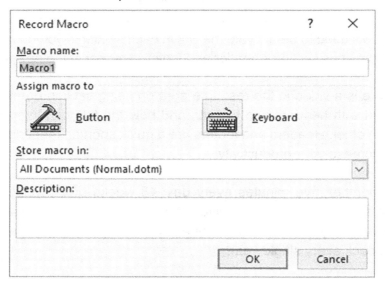

   ✍ Type a suitable macro name, and choose to store it in All Documents (Normal.dotm).
   ✍ Click on OK.
   ✍ Perform the tasks you want recorded.
   ✍ Click on the Stop Recording button.

Macro button

- ⁀ For macros to be activated via a button on the Quick Access Toolbar, right-click on the Quick Access Toolbar in Word.
- ⁀ Click Customise Quick Access Toolbar.
- ⁀ From the Popular Commands drop list, choose Macros.
- ⁀ Locate the required macro, click on it and choose Add.
- ⁀ Click OK and the macro will now be available via a Quick Access toolbar button.

Do anything you can to enable you to work faster, smarter and more efficiently, increasing your productivity, performance and profits.

Tables

The next shortcut is tables. Tables are useful because using tab stops instead can be awkward. A table enables you to lay out your data quickly and easily to look professional and effective. Many people are familiar with tables in Word; they'll go to the Insert tab, insert the table and the number of columns and rows, but there are a range of shortcuts to adding tables. You can also use the inbuilt tools in Word to apply shading, formatting, etc., to make your documents look like you've spent hours on them. You can display tables with no lines in between the rows and columns, so your text looks tabulated and stays lined up neatly. Smarter not harder. Some people call it lazy, I prefer to call it productive! I recommend you look in the resources area on the website for video clips on how to do this. There are some really great shortcuts.

Create a table

There are a number of ways to create a table in Word.

From the ribbon

- ✒ Click Insert tab, click the drop arrow below Table.
- ✒ Select the required number of rows and columns.
- ✒ Click OK.

or

- ✒ Click the Draw Table option and draw the outline of your table, then draw the lines representing columns and rows within the table.

or

- ✒ Choose Excel Spreadsheet – this option allows you to create an Excel "spreadsheet" within your document (note, your system may run slower as a result).

or

- ✒ Choose Quick Table to insert a predefined table layout such as a calendar, etc.

Modify the table

Place the mouse pointer on the line dividing the columns or rows.

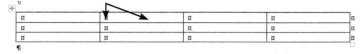

Drag the double-headed arrow to the left/right or up/down to resize the column or row.

To resize a single cell, place the mouse pointer just inside the cell so an arrow displays, click to highlight the cell, then drag the double-headed arrow.

To move the table click the four-way arrow in the top left corner and drag the table to where you want to place it.

Repeat header rows

If your table goes over one page you want to ensure the headings repeat automatically rather than you having to copy and paste them.

> Make sure you are on the table, then from the menu, click Table, Repeat Header Rows.

The heading row will repeat on every page your table continues on.

Create a quick table

Quickly insert a table by typing in +   +   +
The spaces represent the number of spaces in each column.

Page breaks

Another shortcut helps with something that can be frustrating in documents. You can keep pressing the Enter key until you get a page break, and whilst this achieves the purpose, unfortunately, if you need to insert something before the page break it can mess up the document layout. Instead it is better to force a page break. Rather than pressing the Enter key repeatedly until a new page is generated, you can quickly create a page break by holding down the Ctrl key + Enter.

Styles

The next shortcut is Styles. A style is the name Word applies to a range of formatting attributes such as bold, italic, arial, 10 point. Word has a range of inbuilt styles called Heading 1, Heading 2, Heading 3, etc., up to Heading 8, that can be useful to generate consistent headings and to make creating tables of content much easier. Remember the client I mentioned before with the big contract? If they had used the inbuilt styles in Word: Heading 1, Heading 2, Heading 3, for all their headings, 1, 1.1, 1.11, etc., they would have found it so much easier to make changes.

Once you have a style in your document, a heading in style one, for example, you can quickly and easily change the colour of the fonts, the font size, etc., of that particular heading. Or you can easily change the alignment – left aligned, right aligned, centered – for your entire document. You may have gone through an entire document and changed all the headings manually so that they look the same. Styles is a great way of doing this automatically. Start as you mean to continue!

Create your own style

🖑 Select text and make appropriate changes, e.g., bold, italic, purple.

🖑 Click the drop arrow in the Styles grouping.

🖑 Choose Create a Style.

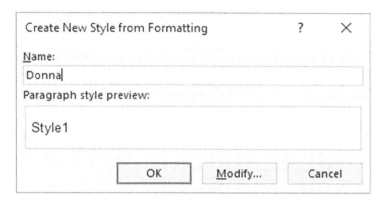

🖑 Give the style a name and click OK.

The style can now be accessed via the Styles grouping.

Apply Word's inbuilt styles to an existing document
🖑 Click on the heading.
🖑 From the Home tab, click on Heading 1 in the Styles grouping.
🖑 Repeat the process for the various headings throughout the document, i.e., for secondary headings, apply Heading 2.

🖑 To generate a table of contents go to References, Table of Contents.
🖑 Click on the desired layout and a table of contents will be created based upon the headings applied in the document.

Table of contents
The final shortcut I want to share with you here – though it is by no means the last one – is Table of Contents. Table of Contents takes Styles and enables you to quickly and easily create a table of

contents. Not only does this look professional in your documents, but it's a great navigation tool for readers to be able to move through a document without having to scroll through the pages. Once you have used the inbuilt styles in Word, i.e., Heading 1, Heading 2, etc., you can then generate a table of contents.

To create a table of contents

- Position the cursor where you want the table of contents to be placed.
- From the References tab, click Table of Contents.
- Choose an inbuilt layout or Insert Table of Contents.
- If you choose Insert Table of Contents, select desired options and click OK.

To update the Table of Contents, position cursor on the table and press F9. A selection can then be made as to whether you wish to update page numbering or the entire table.

Note: You may need to insert additional page breaks in your document so the Table of Content appears on its own page.

If you want to see videos of these shortcuts go to **https://tinyurl.com/DonnaReboot** (coupon code for complimentary registration is REBOOT). Shh ... don't share it with anybody! It's just for the people who've bought the book.

To press Ctrl + Alt + Delete and reboot your productivity with Word, as with Excel, you need to start a conversation. If you work in a team, start having a conversation with your teammates about how you use Word, how other people are using it, what are people leveraging? Share the knowledge.

Style guide

If you work in an organisation, you've probably got a style guide that you need to follow. A style guide is a document that is a foundation for how an organisation wants to present its brand. For example, it might be that the logo always needs to be placed in a particular location and have certain dimensions, typeface and font size, and always be the same colour, or the text might need to be justified.

These form the style guide and show how you as an organisation use applications such as Word, Excel, PowerPoint and email to make sure that the message you're sending to the market is consistent with the branding. If you are in an organisation and have a style guide, you need to be familiar with it. It's useful to learn the shortcuts, but you also need to know how your organisation operates in relation to them.

Starting a conversation with your colleagues enables you to do some knowledge-sharing. If you work for yourself or are in a very small business, this might be with other business owners or entrepreneurs or people in other companies. When you go to a networking event, use it as an opportunity to chat with people and ask them how they do things. People love to talk about what they're doing, and they particularly love it if they know something valuable that other people may not know. So, find out what others know and share your knowledge as well, because that's how we all learn.

In your next meeting, perhaps ask the question: "How do you use Excel or Word? What's working for you? What's not working for you? And what do you need help with?" Simple approaches like these can turbocharge your productivity.

The question that I'm often asked about Excel, Word, and all of the Microsoft Office programs is: "Why would I want to do anything like that? I just go to Google, and Google for answers." Here's what I tell people about Google.

Google is a fantastic tool, and so is YouTube, but it's time-wasting. If you work in an organisation and don't know the answer to something and Google it, firstly, you're looking for something that you may not be able to categorise. You may know vaguely what you're looking for, but the language might not match up with Google's.

So you go searching for something and it could take between five to 45 minutes to find an answer. Then you have to find out how it's going to apply to your workplace, get past the cat videos, and before you know it, there's an hour gone and maybe you haven't even solved your problem. If you have, you go back, apply it and then tick the box and get on with work and don't think to share it with colleagues.

Realistically, if one person in an organisation has a problem or a challenge with something in Word, there will more than likely be other people in the organisation who also have that problem. And they're going to repeat that process every single time. So, starting the conversation could slash through that for you. Why reinvent the wheel?

Recently I worked with a client and we found that people in their team had knowledge that they hadn't shared with others. It wasn't that they had a silo mentality or wanted to protect their jobs, it was more that they'd never been given a forum to be able to share that knowledge. We need to go back to basics, we need to stop being insular and stop just doing stuff, and start talking to each other and sharing our knowledge.

Next we need to plan. Like the contract example I gave earlier, we need to consider our future use. I have a client that processes spreadsheets. They had created a document but not taken the time to maintain it. There were three people in the team and, with some turnover over time, they were using templates that had been left dormant, so the staff were going back to the beginning and doing them again.

You need to consider your future use and make sure you take some time to make some time. It's a bit like time out. We can make the time if it's important to us. After all, if you won a holiday and had to leave on Friday, wouldn't you find a way to finish everything that was outstanding by the end of the week so that you could jump on that plane to go to Hawaii? I know I could. You're able to get the priority items done.

Some things reach a point where, like a house of cards, you can only repair it so much before it falls down, and then takes longer to fix than it would to rebuild it with a better foundation.

So plan!

Think about the use of your document. Are you going to be sharing it with other people?

Are you going to be sharing it with clients? If you are, it needs to have a certain level of professionalism to ensure it fits with your branding.

If you're using it internally, it also needs to be professional, but it also needs to be simplified and automated, which is our next step.

Word has lots of helpful and amazing features that are designed to automate, fast-track and give you back some time. We've already covered a few of them in the shortcuts section: AutoCorrect, Quick Parts, Templates, Styles and Table of Contents. It's important to minimise the amount of manual handling of documents, because the more people that work on documents, the messier they can become, and the more repair work that has to be done by the last person to use the document.

The more you automate your documents, the faster it is to produce the documents; it reduces the risk of errors, reduces re-work, and it means that the information that you're presenting is consistent and structured and you can get back to doing other things, such as talking with customers, liaising with your internal team members, etc. We want to think about simplifying and streamlining for productivity, performance and profit.

Think about your own work. Which of these items would deliver the best result for you? Where is the low-hanging fruit?

What's your priority item in Microsoft Word?

_____

_____

_____

How could it transform your productivity and make your life so much easier?

_____

_____

_____

You may have read all these shortcuts and ideas and thought, "It's really time that I focused on these sorts of things". But you don't have to do everything at once. You've only got to move one thing forward a little at a time and then move on to the next thing. One thing done is better than ten things not done.

Access the resources at **https://tinyurl.com/DonnaReboot** (coupon code for complimentary registration is REBOOT) and work out what your next priority item is and take action now.

CHAPTER 6 – EMAIL

---

*Successful people are simply
those with successful habits.*

BRIAN TRACEY

A study at the University of London found that when you're interrupted by email – when the little box pops up telling you that new mail has arrived – you are interrupted in your thought processes and it can take you 20 minutes to get your mind back to where it was before the interruption.

You'll know this from your own experience. If you think about someone interrupting you whilst you're in the middle of doing something, only to find that when they leave you think to yourself, "Now where was I?". The person who interrupted you has ticked a box, but your productivity is disrupted!

I often hear clients say, "I've got so much to do but I'm not getting stuff done". Rather than measuring your performance by the amount of emails that you send, you need to go back to basics and consider that your work and productivity is measured by your key performance indicators, not the time you walk in the door.

When I first started work, you would come in at a certain time and you'd leave at a certain time. And if you were there for those hours you were considered "productive". In today's world "productivity" seems to be how many emails you send, who you send them to and when you send them. That's not productivity, that's box-ticking. We need a new strategy, a new approach. Too much is done today based upon assumptions. We often assume that how we respond is how everybody else responds. We make assumptions every day, such as assuming that the driver in the next lane will indicate. Do we realise that they might change lanes without indicating?

When I present seminars, workshops or a keynote address on email management, one of the questions I often ask is about responsiveness. How soon should you respond to an email? When I ask that question the answer I inevitably get back is, "It depends". And I reply, "It depends on what?". They say it depends on who it's from, when it comes in, whether they can answer it straightaway; all of these options. My thoughts are it shouldn't matter. If you have your own internal protocol on how quickly you intend to respond, it takes the stress off yourself when you don't assume that when an email comes in you automatically need to respond. That's not realistic, that's not manageable, and that's how you burn yourself out.

Your organisation may have a protocol around using email. When people communicate about how they're going to use email as a tool and understand the expectations, we can start to work smarter and not harder.

If you're in a customer service area and are involved in responding to customers' enquiries, your whole day may involve being responsive to customer requests via email. That's different, of course. But most people in administrative roles or in sales roles need to have a strategy around managing their email. The strategy isn't simply just to hand a new employee a laptop and an email address and say, "Off you go"! If your organisation doesn't have a strategy, it's worth sitting down as a team and discussing how the organisation uses email.

I've often done short in-house workshops with sales teams or finance teams around this topic. Different people in teams often have different assumptions and expectations about how people respond to emails, when they should respond, etc.

With one management team I worked with, one of the challenges was the volume of email they got every day. We got everyone into the boardroom to discuss it. The CEO said, "If you send me an email message and you "cc" me in, I just delete it". The executives were dumbfounded. "Why do you do that?", they asked. "Because if you want me to be across something, put me in the "To" field or simply give me an update in our weekly meetings", replied the CEO.

That worked better for the CEO, but he hadn't communicated it to his team. He was relatively new and his predecessor had wanted to be cc-ed into everything, so nobody wanted to make a mistake or get into trouble.

If you're managing a team, why not sit down and make it clear with them what the rules of engagement are? The team needs to know the framework in which they are operating with email, then they know what they are doing; when they don't know, they have expectations that may not be realistic or attainable.

Another question that regularly comes up is, "What time should you stop sending emails?". One executive had the habit of sending emails on the weekend so that he could get ahead the following week. His team caught on and they started logging in on Sunday evenings to answer emails so they wouldn't be left behind on the Monday.

Realistically, you want your team to be totally disconnected whenever possible on the weekends. There will be exceptions, depending on people's roles; maybe a role requires the odd weekend work, or maybe you are in a role which requires working on weekends on a regular basis. It's about finding the space and the time to have a break, to press Ctrl + Alt + Delete and reboot your productivity. That's what happens when we have time out. But if nobody is told what the expectations are, people do unnecessary things because they think that's what they're supposed to do.

With one sales team I worked with, we asked the question, if you receive an email after 5.00 or 6.00 at night from a colleague, what are the expectations? Rather than being prescriptive and saying, this is what you should do, it's more productive to allow teams to consider what is going to work for them. Realistically, what might be best practice in one organisation might not resonate for you or your team, so you're not going to keep doing it. It's about finding what your best practice is and what's going to make you more productive. It's worthwhile setting time aside and having these conversations in a team. It may seem like a waste of time, but less stressed, more productive people with clear expectations are going to increase productivity and turbocharge what you can get through.

Now let's look at some shortcuts or "hacks" to work smarter and not harder with email. Not all are application-specific but some of them are.

Working with a global client, one of the first things we did was talk about email and linking it back to key performance indicators. Whether you use Outlook or something else, set it up so that it opens in your calendar and not in your inbox. This means when you start up your email you're reminded of what *your* priorities are for the day rather than being driven by other people's priorities.

Set Outlook up to open in Calendar view

A way to remind you of your appointments and activities and keep you productive is to set Outlook to open in the calendar rather than straight into your inbox, which is the default. This enables you to focus on what your priorities are for the day, rather than be drawn immediately into requests from others.

Set Outlook to open in the calendar

- ⍟ Click the File tab.
- ⍟ Choose Options.

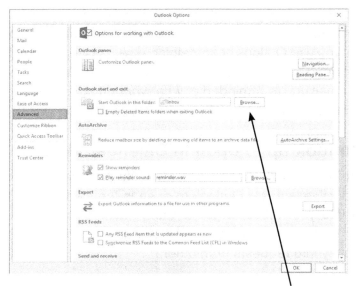

🖰 Click the Advanced tab and choose Browse beside "Start Outlook in this folder".

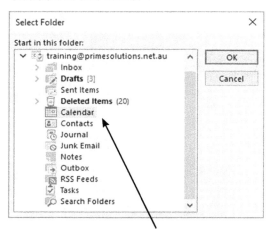

🖰 Click Calendar, then click OK.

🖰 Click OK again.

🖰 Close down Outlook and reopen it.

It will now open in the Calendar view.

In the resource library at **https://tinyurl.com/DonnaReboot** (coupon code for complimentary registration is REBOOT) you'll find a video that shows you how you can set up Outlook so it automatically starts in your calendar every single time. This simple thing can save you so much time every day because it reminds you of what your priorities are.

The next shortcut is to turn off any notification items. If you're accessing your email on a smartphone, switch off the notifications unless there's a specific reason why you need to receive emails. If you are expecting an email from someone, you can check your inbox, but the constant interruption of notifications impacts on your productivity. As mentioned earlier, studies show that every time you're interrupted it takes you 20 minutes to get back to where you were. So if you're interrupted five times in an hour, each time it's going to take you 20 minutes to get back to where you were. That is a lot of lost productivity. Turn off the notifications!

Turn off email notification

By default, when you have Outlook open, when new emails arrive in your inbox, a notification box pops up and a "ping" sound plays to let you know you have new mail. Whilst this can be useful, it can also be a distraction when you are trying to focus on something.

To switch it off

🖰 From the Menu, choose File, then Options.

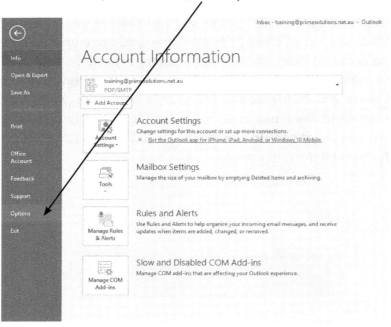

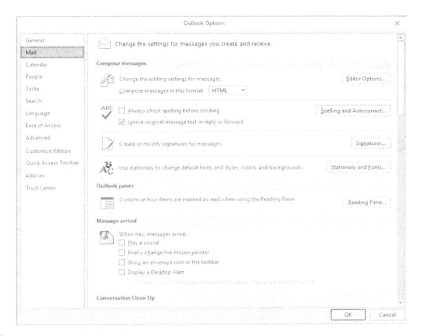

⌐ Click the Mail tab.
⌐ Remove all the ticks from the Message arrival checkboxes.
⌐ Click OK.

For this to take effect, you need to close Outlook then reopen it.

In the resource area at **https://tinyurl.com/DonnaReboot** (coupon code for complimentary registration is REBOOT) you'll find a video on how to do this in Microsoft Outlook.

The next shortcut is to create some draft templates. There are a couple of ways to create templates for emails that you send out on a regular basis. The first is by creating an email message with all the relevant data in it, but with elements left blank to enable you to fill in the relevant details each time.

Repetitive messages

Sometimes you have a message that you need to resend to several people on different occasions. The information is mostly the same each time, but the names are different and you need to send them at different times. Save some time by creating some pre-prepared drafts.

Create drafts

- Create the message and press the Escape key on the keyboard to save it as a draft.
- Use the Ctrl key to drag the message and duplicate it. (Make sure you release the mouse, then the Ctrl key or nothing will happen!)
- Once you have done it once you can select 2 or 3 and duplicate them again.
- When you need to send them, go to your Drafts folder, double-click to open the message, complete the details and press Send.

Another way to do this is to use a signature, which means you can quickly click on something and it will drop in a key piece of text. It is useful for emails that you want to send, such as setting some appointments.

Signatures

You can set up various signatures in Outlook to quickly drop closing text into emails, such as your contact details. For instance, you may want to use one signature when responding to internal messages and another, with more information such as a phone number, for external messages. In Outlook you can set up a default or move between various signatures.

Create a signature

🖰 Click the File tab and choose Options.

🖰 Click the Mail tab.

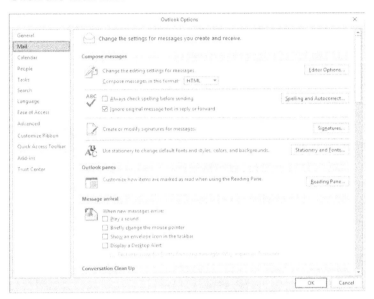

🖰 Click Signatures.

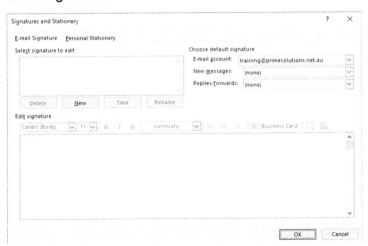

🖱 Create a new signature by clicking New.

🖱 Give the signature a name.

🖱 Click OK.

🖱 Type the signature in the Edit signature box.

Note: Leave a blank line to allow for the email message to be written above.

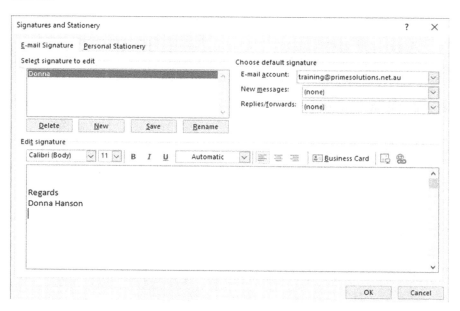

Once you have finished, you can choose whether you want this signature to display in all new messages and replies and forwarded
messages by selecting the signature from the drop list.

Alternatively, if you prefer to activate the signature manually, leave
both options with (none) displaying.

ᐟᕀ Click OK and OK again.

Activate the signature

ᐟᕀ Click the new email icon.

ᐟᕀ Click in the main message area then click the Signature drop
arrow to choose the desired signature to place in your email
message.

Note: If you have multiple signatures, you can toggle between them
to place the appropriate one in the message.

The next shortcut is, rather than send attachments, to send a
hyperlink. This is particularly relevant where you're emailing internally
to your colleagues.

One of the biggest challenges for IT departments is the speed and ease that user mailbox size can be depleted through large attachments. For example, say I email you a document that's three megabytes in size. I email it and I now have a three-megabyte attachment sitting in my sent items. Who goes to their sent items and deletes those attachments once they've sent them? Most people don't. The attachments just sit there. The person who receives it gets the attachment, and it's three megabytes in their mailbox. They may save it to a network drive, but the likelihood is they just leave it attached to their message.

If they then make some edits, reattach the file and send it back to you, they've now got a 3MB file not only in their inbox, but also in their sent items, and there are now another three megabytes in your inbox.

A smarter way to do this, and to save time and reduce the chance of ending up with multiple versions of the same document, is to use a hyperlink. Where possible, and if your colleagues have access to documents in a shared drive, send a hyperlink instead of an attachment. Not only does it eliminate the problem of file size (because there isn't an attachment), it also means you're both accessing the same document, which is useful if you're both editing. You just need to make sure that both of you have access to that drive location.

If you're sending something externally, if it's a downloadable document that you want them to be able to access, attach it as a PDF. This locks everything up and reduces the file size, enabling you to take up less space. Alternatively, if it's to someone external, rather than attaching something, can it be found on a website? Can you direct the person to a hyperlink on your company website for them to download that piece of information so you're not clogging up their inbox?

If you send a 5MB file to someone, you may find that their organisation only allows attachments up to 3MB to be received for security
reasons. You may find that they never actually receive the file, so if
you want to send a larger file, it's worthwhile checking to see if there
are limits. I've seen people try to send 50 or 100MB files only to
find that the intended recipient never actually received them, and in
some cases they don't even get notified.

Hyperlinks

Hyperlinks in Outlook are an alternative to sending an attachment
and offer a few benefits. Hyperlinks link to a single file in a shared
location. To use a hyperlink in an email the receiver must be able to
access either the network drive where the file is stored, the online
drive location or a link to a website. If the intended receiver cannot
access the location where a file is stored, an alternative method of
sharing needs to be used.

A hyperlink is better than an attachment for a couple of reasons.
Firstly, it reduces the likelihood of multiple copies of the same file.
Secondly, it reduces mailbox space as it isn't an attachment; it's
simply a "shortcut" to where the attachment is stored.

🖱 To insert a hyperlink you need to be in the body of a mail
message.

🖱 Click the Insert tab and choose Link.

🖱 Click Insert Link.

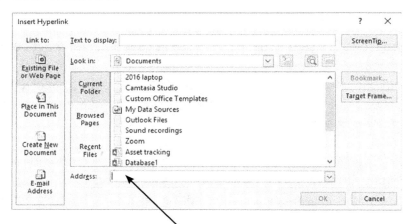

🖱 Locate the desired file or type in a web address, if appropriate.
🖱 Click OK.

The hyperlink is now inserted in the body of the email message.

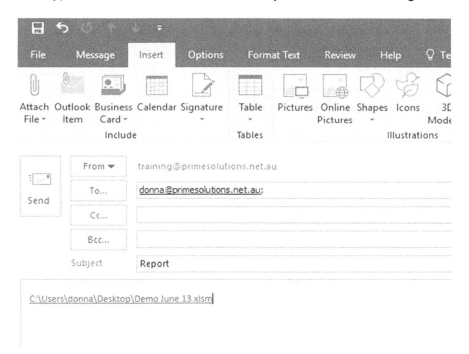

🖱 Complete the message and send.

Brain dump

The next shortcut is doing a regular "brain dump". This is a bit like a to-do list. Basically, it enables you to brain dump everything you need to do that's in your inbox into your calendar. Everything that goes into your calendar when you're allocating time is more likely to be done than if you have a to-do list that sits somewhere externally, as you madly tick items off the to-do list.

If you have a monthly report that you need to prepare, and you know that every time you do it it's going to take you an hour, why not plot that into your calendar every single month? That way, you've dedicated the time. If someone needs to book an appointment with you, they can see that that time isn't available. So put as much as you can in the calendar, because what gets scheduled gets done. As Stephen Covey, author of *The 7 Habits of Highly Successful People* says, "The key is not to prioritize what's on your schedule but to schedule your priorities".

The next shortcut is to delay sending. A common occurrence is sending an email message and then realising you forgot to attach the attachment. Whilst some programs pick up the fact you've forgotten to attach a file, based on a keyword in the email such as file or attachment, some don't, and there's nothing more embarrassing or unprofessional. Or perhaps you send an email and five minutes later you realise that you have to send another one because you missed a key piece of information. Switching on the delay sending option in your email is perfect for this.

When I suggest it, people often say, "But I don't want that to happen, I want the email to go straightaway". However, if you delay sending a message for ten minutes, you'll find most of the time that ten minutes is the magic number by which you realise, a) you've forgotten to attach a file, or b) you need to say something else. Anything longer than ten minutes and you can potentially send another email or pick up the phone and talk to somebody.

By delaying sending it means that the item sits in your outbox for ten minutes. In the event that you do want to force the email to go, you can easily click on the send/receive button to force it to go (provided you've got internet access, of course!).

Delay sending

🖰 From the File tab, choose Options and then click the Advanced tab.

🖰 Scroll down to "Send and receive".

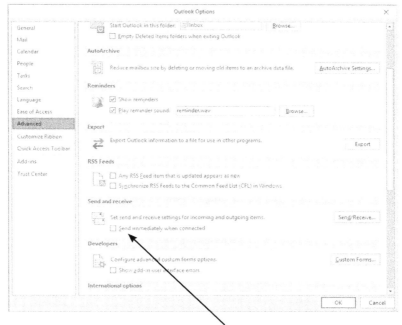

🖰 Remove the tick from the "Send immediately when connected" checkbox.

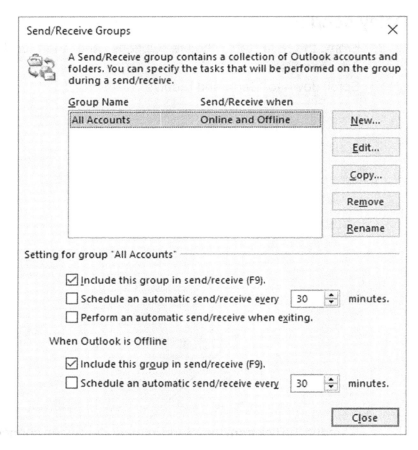

✎ Change the "Schedule an automatic send/receive" option to five or ten minutes and click Close.

Note: At any time you can override it by pressing the F9 key on the keyboard or clicking the send and receive icon on the Home tab or the Quick Access Toolbar.

Rules

The next shortcut is called rules, which are amazing! Rules are like having your own personal assistant in your email program, and can apply to messages that you send or receive. They enable you to filter emails that aren't urgent, but are still important, into other folders for you to review later. Some examples of useful rules include redirecting newsletters to folders to read later, redirecting applications for jobs to a job folder to review later, moving messages sent or received from a client into that client's folder, etc.

It is useful if you regularly receive emails, such as newsletters, that you like to read but you don't want them to go into your inbox with everything else. You can create a rule that enables you to move messages that meet certain criteria to a folder to be read later, so they file automatically and you go to them when you have time to read them.

You'll know when there are new items in a folder, because a number in brackets will appear beside it in bold, letting you know that there is new data in there.

Rules are great to explore to fast-track your productivity, because they also enable you to automatically file your sent items. For example, you can set up a rule that when you send an email to anybody with, for example, "abc.com.au" in the domain name, it will automatically go into the ABC client folder. All of this will help you to leverage and press Ctrl + Alt + Delete and reboot your productivity.

Click the File tab and choose the Manage Rules & Alerts button on the Info tab.

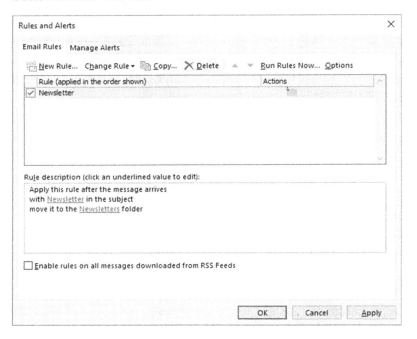

🖱 Choose New Rule.

🖱 In the Stay Organized section, click on "Move messages from someone to a folder".

🖱 In the Step 2 window, click the links to choose the people or public group required and where to move the message to.

🖱 Click Next.

🖱 Keep clicking through Next and choose the appropriate options.

🖱 Click Finish when complete.

Note: Use the "Start from a blank rule" option if you want to create a rule that applies to messages you have sent.

The next shortcut is to go offline. I know you're probably thinking to yourself, "That's counterintuitive. This chapter is on email!". As I've said before, sometimes we need to take time to make time. We need to go offline to be more effective. This is particularly useful when you've got tasks that require your complete concentration. Closing down email when you need to focus on a detailed Excel spreadsheet or a proposal that requires your complete and undivided attention is very important.

By focusing your attention totally on something, you'll find your productivity is actually increased. You get things done quicker when you're not distracted. As the University of London statistics suggest, the fewer interruptions you have, the more productive you are. I'm sure you've had the kind of day where you've gone home at night and told your partner, "Wow! What a day I've had today. I got so much done!". Compared to the day you go home and say, "Oh, I got nothing done; I felt like all I did all day is send and receive emails". You need to press Ctrl + Alt + Delete and reboot your productivity – go offline to work smarter and not harder.

The next shortcut with email is to create your own personal responsiveness framework. What do I mean by this?

I mean you need to work out what your frame of reference is about how quickly you will respond to email from other people. You might be guided by internal protocols or what you do as a team, but if you can mentally say to yourself, "In 99.9% of cases I must respond to emails within 24 hours", work with that. You might find your framework is two days. It's not up to me to say what is best practice, because best practice is what's best for you. You need to find something that you can maintain. So what is it? How quickly should you respond? What is your own personal responsiveness framework which you can apply every time you see an email come in.

Here's a bonus tip around email management.

The bonus tip is a strategy called Email SOS. Email SOS is the foundation of a keynote address that I present.

Email SOS

The first S in Email SOS stands for:

Stop sending

Think about the 'law of attraction', that is, what you get out is what you get back. If you send lots of emails, you probably get lots back. If you stop sending emails, you'll stop getting so many back.

How do you get your work done if you're not sending emails? What about picking up the phone? What about making a list of all the things you need to talk to someone about, then pick up the phone or arrange a meeting to discuss them, rather than dealing with bits and pieces here and there? What often happens is when you shoot an email to somebody, you're usually not the only person asking a question of that person. If your email is vague or has too much information or too many questions, it often leaves the receiver not knowing what to do, so they do nothing. Stop sending and you'll stop getting a lot of email back!

The O in Email SOS stands for:

Offline

We've covered this a little already. It's OK to switch off your email when you're in a meeting. It's OK not to have your email open when you're not in for the day or you're home sick. If you're home sick, you shouldn't really care.

I often say to people in my presentations, we don't need to be online or accessible by email 24/7. If something happens that is an emergency, we have provided an emergency phone number. I'm yet to see a form where they ask for an emergency email address. The world is not going to end if you close your email down and go offline to enable you to focus.

The final S in Email SOS is to:

Stop others sending

You might be saying to yourself, "How do I get other people to stop sending me emails?". This can be quite easy; it's just a matter of communication. For example, when I worked with a school, one of the greatest challenges was the volume of email they received. When we explored this and asked staff questions in a pre-session survey, we found that a large volume of their email was a result of internal emails, with people simply saying things like "thank you", or because they were using "reply all". If we have a conversation with our colleagues we can get into the habit of not doing these things. We can say something like, "OK, from now on let's not say thank you to emails or choose reply all. Let's just agree that we're grateful for anything that we've received and that we are not here to create extra work for each other!". Once there is agreement, it can reduce the volume of email people receive by as much as 30%. I've seen it happen with organisations that I've worked with.

So there's your bonus tip! Email SOS. Stop sending, go offline, stop others sending.

Want some help getting your team or organisation
on the same page with email?
Visit **www.donnahanson.com** to find out how Donna helps.

We make a lot of assumptions, and I've shared some examples around sales teams and executives. You need to find out what's right for you, your team or your organisation. You need to create your own email protocol and organisational protocols. You need to be clear in your conversations and not consider emails as an opportunity to just go "blah blah" and blurt out everything that you want to share! We're concise, we're targeted and we're considered when we create a letter, so why aren't we the same with email? Make the commitment by having the conversation with your team or your colleagues or your organisation. Consider the following:

Are you going to stop sending thank yous and reply alls?

Create a consensus and have a shared understanding, because when everyone knows what the expectations are they can all move in the same direction.

Once you've had the discussion you can plan your own personal protocol.

You can use some of the shortcuts to set up repetitive emails. You can create rules to set up a filing structure that makes it easier to find and search for information in your mailbox. You can come up with some inbox strategies that are going to enable you to work smarter and not harder, and really leverage technology, and press Ctrl + Alt + Delete to reboot your productivity.

In other chapters we've talked about automation.

One of the shortcuts is to automate in Outlook, but you also need to consider time. Allocating times for recurring meetings so that things are blocked into your calendar. Synchronising your smartphone with your Outlook calendar so that all your appointments cross over and display on all of your devices. Adding items in your mailbox to your contacts so that any time you're on the road and running late for a meeting, you know you can easily access the contact details of the person in your smartphone or your tablet. Quickly and easily knowing how to create shortcuts for tasks and calendar entries as well as rules are fantastic ways to automate your activities in Outlook.

Set expectations and don't make assumptions! Use email as the tool it was meant to be: to confirm conversations and *not be the actual conversation*, wherever possible. Think of your own work and make that a priority. Remember, five minutes saved on one task in one day equates to two-and-a-half days per year.

We've looked at email, we've looked at Word and we've looked at Excel. So if you could save five minutes in those three areas, already you've saved seven-and-a-half days a year.

So now is the time to press Ctrl + Alt + Delete and reboot your productivity.

Don't forget to access the resources at **https://tinyurl.com/ DonnaReboot** (coupon code for complimentary registration is REBOOT).

CHAPTER 7 – SOCIAL MEDIA

Lost time is never found again.

BENJAMIN FRANKLIN

When I speak at conferences or industry events, I'm often asked about social media. People ask, "What social media should I be on? What's right for me? How do I monetise it and how do I get a return on investment of my time?".

Social media is important for a number of reasons, and I think we need to be clear about what our purpose is. I see it as positioning, educating, establishing your expertise, or reminding your audience of what you do or who you are.

I believe you need to have a planned approach, but it depends on your circumstances. If you're in an organisation, you need to be guided by its social media policy or strategy. Why? You need to consider that anything you post while you're employed by an organisation is, in effect, a representation of their brand. Legal cases have ended with people losing their jobs because they have posted negative comments about the company they work for. If you are a solo entrepreneur, you have a bit more flexibility.

What's right for you depends on your role, but you need to consider both personal and professional aspects. I'm very mindful of my social media footprint, so for me it's generally all business-related, and I spend a little more time on LinkedIn.

There are four main social media platforms to discuss.

The first is Facebook, which is considered mostly "business-to-consumer". It's also a great way of creating a personal connection with some of your strategic contacts in large organisations. The value of groups on Facebook can't be underestimated. I'm part of a group of professional women speakers around the world, which I find extremely valuable. I don't tend to use Facebook a lot otherwise, because most of my market is primarily business-to-business.

The second is LinkedIn. LinkedIn is considered the "business-to-business" platform. It's about sharing and educating and learning and positioning. It's a great opportunity for people to look for potential candidates for jobs, and for you to find connections to people who might be able to introduce you to the decision-makers that you need.

The third is Instagram. This tends to be a bit more personal, although if you have a product that you want to sell, business-to-consumer, Instagram could be for you. It's very image-focused, so it includes a lot of pictures.

Twitter, the fourth platform, can be both personal and business, business-to-consumer or business-to-business. It's text-based and it uses 140 characters.

The key thing to do to Ctrl + Alt + Delete and reboot your productivity is to find the one or two platforms that work best for you. Don't try to get across them all!

Think about where the people you liaise with are more likely to go.

What are you wanting to do? Are you looking to position yourself? Are you looking to educate your audience so they know, like and trust you and will want to buy from you?

What is your purpose and who is your market? That will help you identify the platforms you need to be on.

Once you've found your platforms, focus on them, but be mindful of business versus personal. Remember that recruitment companies and potential employers could be looking at your profiles and may make decisions about you as a potential employee based on your Facebook posts. So make sure your social media represents you well.

If you want to know more about social media, there are some great resources. I highly recommend two books by personal branding and LinkedIn specialist, Jane Anderson, *Connect*, and *Expert to Influencer*. Both books offer valuable insights on LinkedIn and how to use it strategically.

There are a lot of programs online to help you maximise Facebook for selling and other options. Try your local library, visit YouTube or take a look at programs on Udemy.com or Lynda.com.

Pressing Ctrl + Alt + Delete and rebooting your productivity with social media is to be mindful that it can be very easy for you to get absorbed in social media and feel like you're doing marketing or "sales-ey" sort of stuff, but what can happen is that your day evaporates!

If you have decided to go down the social media path and have a couple of particular platforms that you want to focus on, I recommend being strategic about what you do. By using programs such as Hootsuite, Buffer or Meet Edgar, you can queue up three or four different posts for a particular week around your area of expertise, and automate your posting every day or so. This enables your posts to be automated and to go out in sequence, rather than you having to remember to post, or your posts being sent randomly.

You can do the same with LinkedIn, Instagram and Twitter as well. The key thing to remember is that it's easy to spend time on social media, but does it align to your strategic goals?

So, what would happen if you were able to get five minutes more productivity each day, week or month? That would give you two-and-a-half day's minimum extra time per year. But what else are you capable of?

PART 3 – IMPLEMENTATION

The secret of getting ahead is getting started.

MARK TWAIN

Congratulations on investing in this book and for reading so far. Now is the time where the rubber hits the road! You need to take some action if you really, truly do want to press Ctrl + Alt + Delete and reboot your productivity.

The next few chapters discuss how you can go about creating your own strategy and implement processes by which you reboot how you perceive, use, learn and leverage technology.

For additional resources, visit **https://tinyurl.com/DonnaReboot** (coupon code for complimentary registration is REBOOT).

Take time to make time. Make this investment of your time and your money worthwhile so you can press Ctrl + Alt + Delete and reboot your productivity.

Turn the page to get started!

CHAPTER 8 – TIME

You can always make another dollar, but you cannot make another minute.

ALAN WEISS

Now we get to the implementation part. Let's look at how we take the next step and implement what we've learned.

Richard Branson, Bill Gates, Oprah Winfrey and even Kim Kardashian all have one thing in common. What's that? All of them have 24 hours in a day; nothing more, nothing less. Each of these people is strategic in how they choose to use time. All are exceptional at getting things done to move their brands and their business forward with more speed than the majority of everyday people.

Sure, being a celebrity might help, but the key to getting things done is knowing right now what the right things to get done are. I believe that pressing Ctrl + Alt + Delete and rebooting your productivity is as simple as doing the right things at the right time to produce the right results. We need to take time to make time. It doesn't take hours, but here's a simple plan to get started and to give you back more time.

Here's the four-step process to giving you back more time.

Identify the low-hanging fruit

We've learnt more about several applications. Whether you're increasing your productivity so you can get more done, or whether you are in sales and you're increasing your productivity with technology so you can talk to more clients, you need to identify what the low-hanging fruit is. What are the easiest, simplest, quickest ways for you to get some runs on the board to quickly increase your productivity?

_____

_____

_____

It might be as simple as creating some templates or signatures in your email that mean that some of those repetitive emails you send are far easier and quicker to create. It might be being strategic about when you're going to deal with emails, or switching off your email notification. Maybe it's in Excel where you can commit to learning a few tips and shortcuts to enable you to slash through some of the repetitive stuff that you do. Or maybe it's in Word where you can create documents that are consistent in their structure.

Identify the low-hanging fruit, and I recommend that you pick only one or two items. Even if it's just one or two things in email, focus on the area that's causing you the most annoyance or costing you the most time in productivity and get some time back there. One thing at a time. We need to walk, putting one foot in front of the other.

Pressing Ctrl + Alt + Delete and rebooting your productivity is about doing one thing at a time. Implement that into your daily activities and before you know it that five minutes turns into two-and-a-half days a year.

Consider what needs to be done

Using Microsoft Excel as an example, I worked with someone in finance who was doing a budget, but every time they sent the budget to other departments, they would change names and delete items that didn't apply to them. We combined a few simple tools and protected the worksheet and made items green to let staff know where they needed to input their data. We simplified the documentation. What needs to be done with your one area to get you kickstarted?

Create an action plan

What steps are you going to take right now that will enable you to move forward and achieve this goal? If you've ever done goalsetting 101, you know you need to have an action plan and you need a timeline. You've identified your tasks and worked out what needs to be done. What are the key steps and when are you going to do them by?

If you can do this, you are in total control. Remember, for every five minutes you save that's two-and-a-half days a year. That means less stress, increased productivity, more time to do the things you love, to be with your family, with your friends, doing the fun activities, or simply just having some downtime.

Implement and reward

Allocate a realistic amount of time to implement. And once you do, consider rewarding yourself in some way. If your task is a big one, you might want to give yourself something, or simply acknowledge that extra time that you've got and enjoy it. If it didn't work out exactly as you planned this time, do a debrief and work out what went wrong and how can you do it better next time.

CHAPTER 9 – TOOLS

I think it is fair to say that personal computers have become the most empowering tool we've ever created. They're tools of communication, they're tools of creativity, and they can be shaped by their user.

BILL GATES

It's pretty hard to nail a screw into a block of wood or put a square peg into a round hole, but that's what we often try to do with our technology tools in today's business world. It's important to find the right tool for the right job.

In this book we've looked at Excel, Word, email and social media. Let's now have a look at what tools we can use to move these things forward.

The first tool to think about is social media. Which platforms will you use? Which ones are right for your market? Which ones come easy to you and don't seem like drudgery to have to do on a regular basis? Let's be honest; if you feel like it's a drudge, you're less likely to do it. You want it to be a medium that you already enjoy and engage with because of its style and because of your other interactions, etc. So, work out which social media platforms you're going to use.

_____

_____

_____

Next, ask yourself, what do you need to know or find out to be able to plan or do? Are you going to post random things as they come up or are you going to be strategic? Is it about educating your customers and contacts about what you do, so that they will remember you when they come to need whatever product or service it is that you sell? Having clarity about what your purpose is and what you're going to use is important.

I provide a range of productivity pointers – one tip, trick or shortcut once a month in Word or Excel, to all the contacts on my mailing list. You can sign up at one of my websites: www.donnahanson.com.au, or the www.primesolutions.net.au.

I'm building expertise, building trust and engaging with my audience, without having to ring everyone every month. What could you be doing?

_____

_____

_____

The second tool that we use is email. What are some of the things you need to activate or start doing with email? Is it changing Outlook so that when it opens, it does so in the calendar? Is it creating some new signatures? Is it about creating a rule that applies to your sent items to fast-track the filing of emails?

How can these email tools benefit you? How can they make your life easier? Do you need to consider other colleagues or your organisation? Should we be operating as a team?

_____

_____

_____

The third tool is Microsoft Word. Where's the low-hanging fruit? Has someone in your organisation already done something that you could use? Have they already created a table for something that's similar to what you do? Have they already got a document, a range of documents or reports set up that you could leverage off?

Two heads are better than one, so why create something from scratch when there could be someone in your organisation who has already done the hard work for you; you just need to find them. The best way to do this is to start upskilling your colleagues. Start talking to people about how they can work smarter and not harder with the Microsoft Office programs, because if you can all help each other, then you won't be going to Google and searching for something that will take time and end up with you getting distracted by cat videos! You may solve your problem, but you just tick the box and move on. If one person in an organisation needs the answer to a question, the likelihood is there'll be at least another one or two, which means that several people are losing time you can't get back. It's unproductive time.

The important thing to remember is you need to build your skills on an ongoing basis. The easiest way to do this is to have regular forums or focus groups within your organisation, maybe once a month, to chat about how you're using Word: what's working, what's not working, and what's your greatest challenge? We need to start talking about it.

The fourth tool is Microsoft Excel. With Excel, ask yourself, where does the biggest risk lie in your spreadsheets? A lot of the time our spreadsheets are open, because we're referencing them, rather than locking the data and minimising risk. Often we assume that the data in Excel is correct – and we shouldn't!

You also need to work out which spreadsheets are wasting your time and if you need to automate. Automation elements such as drop lists or macros or sort and filter and conditional formatting are designed to make your life easier. Ask yourself why you need to do something, what you need to do, and then how you're going to do it. Does a shortcut exist? Can something be fast-tracked?

_____

_____

_____

Read through the Excel chapter again and think about the things that you could apply straightaway. How can you reduce risk and reduce the amount of manual labour? It's worthwhile investing time in Excel, because a lot of decisions are made using data in Excel spreadsheets, and it can be an area where a lot of people struggle.

_____

_____

_____

Don't forget, you can visit the resources at **https://tinyurl.com/DonnaReboot** (coupon code for complimentary registration is REBOOT) to learn more.

CHAPTER 10 – TEAM

Great things in business are never done by one person. They're done by a team of people.

STEVE JOBS

Whether you're in a company, a school, or are self-employed, we all need a team. We need supporters to share, learn, do and grow.

At the start of Part 1, I mentioned former LA Lakers coach, Pat Riley's quote, "If you're not getting better, you're getting worse". The only way to get better is by investing in yourself and learning more. However, you can learn a lot faster when you're working as a team.

Napoleon Hill wrote the book, *Think and Grow Rich*. It focused on the importance of masterminds, the importance of sharing knowledge with like-minded people who are on the same journey as you.

So let's talk first about sharing. How do you share information?

If you work for a company, it might be in the team that you're a part of. If you don't already have one, you could create a team in your organisation. Depending on the access to resources in your company, you may be able to find two, three or four people in other areas of the business with whom you can get together on a regular basis and discuss how you do things.

What could you fast-track or streamline? How, as a group, can you reboot your own personal productivity and your team members'? You can become productivity champions for Ctrl + Alt + Delete, reboot your productivity. You can talk to each other about what you know, what's working for you with technology, what's not working for you, and what you need help with?

When you enter a meeting with these pieces of information, you can start sharing your ideas and you can help one another work smarter and not harder. If you're in a school you can do the same thing with colleagues; you can get together for professional development days. If you're self-employed, use the Napoleon Hill process and find some like-minded people that you can share and learn and grow from and with. We learn by being in groups and sharing ideas and discussing opportunities, and you'll be surprised how much others know.

We need to get back to basics, talking to each other and building relationships beyond the technology. I believe that just because we've got the technology doesn't mean we need to use it. In fact, I think we need to start using it less to get more done.

Recently, I had a conversation with a colleague in which I mentioned some websites and how I'd found that the resources on those websites had leveraged my productivity through outsourcing. This colleague didn't know about these resources, and I was able to introduce her to some interesting insights and connections that will now enable her to work smarter and not harder.

I don't want to keep everything I know to myself. My goal is to help people learn and grow, because I've had to learn and grow, and I want to pass it on so that you can press Ctrl + Alt + Delete and reboot your productivity, get things done and get on with your life.

It's about creating a team internally or externally, depending upon your circumstances. As a team, you need to work out how you can grow, how you can leverage your ideas and how you can support each other. It might be sharing your knowledge, it might be expanding and growing your knowledge, it might be problem-solving. We need to get out of the silos and we need to start sharing our knowledge so that we can all stop working so much and start focusing on living life.

I recommend that you go to the resource library which can be found at **https://tinyurl.com/DonnaReboot** (coupon code for complimentary registration is REBOOT) to learn more and consolidate what we have covered in this book. My goal is to help you press Ctrl + Alt + Delete and reboot your productivity.

FINAL NOTES

Thanks for reading *Ctrl + Alt + Delete - Reboot Your Productivity*.

I hope in these pages I've generated ideas and strategies that will enable you to focus on the "low-hanging fruit", the chance for you to quickly and easily convert opportunities to create productivity into real value for you.

Everything in the book is current at the time of printing, however, technology changes every day. I recommend you go to the resource library and sign up to make sure you can access a whole range of resources on what we've covered in *Ctrl + Alt + Delete - Reboot Your Productivity*.

You can also take the next step. At the back of this book you'll find a range of in-house presentations and consultation details, or you may be interested in an individual or organisational coaching program. I'd love to help you on your journey to reboot your productivity.

Now it's over to you. As Lewis Carroll said in *Alice in Wonderland*, "If you don't know where you're going, any road will get you there".

Make the decision today to press Ctrl + Alt + Delete and reboot your productivity and your life. I'd love to hear your thoughts, inspirations and ideas, and learn what's helped you and how much time it has saved you. Email me at Donna@donnahanson.com.au.

Yours in productivity and success!

Donna Hanson

THANK YOU

This book is dedicated to my family, friends and colleagues. You know you are important to me.

I also want to thank my mastermind groups. You know who you are and how amazing it is to be soaring with such eagles!

Thanks also to Professional Speakers Australia and National Speakers Association (USA) and Power Women.

I also dedicate this book to the clients I've worked with around the globe, from Australia, UK, Japan, Singapore, Cook Islands and New Zealand.

I thank you for the privilege of allowing me to help you increase your productivity and reduce your stress and for all the lessons you have taught me, and continue to teach me.

I thank you for being brave and choosing options that sit outside the norm as far as your learning is concerned.

REFERENCES

Jane E. Anderson, *Expert to Influencer*, 2017, www.janeandersonspeaks.com

Jane E. Anderson and Kylie Chown, *Connect: Leverage your LinkedIn Profile for Business Growth and Lead Generation in Less Than 7 Minutes per Day*, 2015, www.janeandersonspeaks.com

Stephen Gandel, *Damn Excel! How the 'most important software application of all time' is ruining the world*, 17 April 2013, http://fortune.com/2013/04/17/damn-excel-how-the-most-important-software-application-of-all-time-is-ruining-the-world

Medibank Private, *The Cost of Workplace Stress in Australia*, August 2008, www.medibank.com.au/client/documents/pdfs/the-cost-of-workplace-stress.pdf

Microsoft Canada, *Attention spans*, 2015, www.scribd.com/document/265348695/Microsoft-Attention-Spans-Research-Report

Alan Weiss, *Million Dollar Consultant*, 5th ed., 2016, www.alanweiss.com/store/books/million-dollar-consulting

ABOUT DONNA

Donna Hanson is a Certified Speaking Professional (CSP), the only globally recognised certification for professional speakers. Donna has worked in the adult learning and development space for over 20 years and holds a Bachelor of Training and Development degree.

She is one of only five people outside the USA accredited to deliver email productivity training to Microsoft's own staff.

Donna regularly presents at conferences and organisational events and is regularly called upon by media to comment on technology-related topics. She has presented in Australia, New Zealand, Singapore, Japan, USA, United Kingdom and the Cook Islands.

Donna is passionate about getting people off their computers, so they can get on with what matters: enjoying life and spending time with the people they love.

WANT MORE?

Want more books for your team?

Order online at www.donnahanson.com.au/shop, or contact our office on +61 3 9457 4745, or email your enquiry to enquiries@donnahanson.com.au.

Want to have Donna present for your team, organisation or upcoming conference?

Donna regularly presents at conferences, in-house professional development days, retreats and team-building days.

Her current keynote lectures topics can be found at www.donnahanson.com and include:

- Email
- Productivity
- Microsoft Word
- Microsoft Excel
- Teamwork and Collaboration
- Planning

COACHING –
INDIVIDUAL AND TEAM

Donna is available to a limited number of organisations as a productivity expert and coach. For more details on how Donna can help, visit www.donnahanson.com.au

ONLINE RESOURCES

Complimentary access to online resources valued at $147

Ctrl + Alt + Delete – Reboot Your Productivity has a companion range of online video resources which are available to you with the purchase of this book at no additional cost.

To gain access to the resources you need to "purchase" the course.

Use the coupon code REBOOT (all capitals) at checkout (this will reduce the price from $147 to $0).

Once you have "purchased" the course with your coupon code, you will be able to log in with the account details you created. You will also receive an email with a link to the login site which will enable you to access the resources anytime.

The link to go straight to the Reboot site is **https://tinyurl.com/ DonnaReboot** (coupon code for complimentary registration is REBOOT)

Should you have any problems accessing the resources, please email enquiries@donnahanson.com.au.

Donna also has a range of online courses and resources that can be accessed via www.donnahanson.com.au/shop

Want to connect with Donna on social media?

Send an invitation and make sure you mention the book.

Facebook – donna.hanson.311
LinkedIn – donnahansoncsp
Twitter – @donna_hanson
Pinterest – donnahanson5